The Many Faces of Science

THE MANY FACES OF SCIENCE

An Introduction to Scientists, Values, and Society

Leslie Stevenson
University of St. Andrews
Fife, Scotland

Henry Byerly
University of Arizona
Tucson, Arizona

Westview Press
Boulder · San Francisco · Oxford

Copyright © 1995 by Westview Press, Inc.

Published in 1995 in the United States of America by Westview Press, Inc., 5500 Central Avenue, Boulder, Colorado 80301-2877, and in the United Kingdom by Westview Press, 36 Lonsdale Road, Summertown, Oxford OX2 7EW

Library of Congress Cataloging-in-Publication Data
Stevenson, Leslie Forster.
    The many faces of science : an introduction to scientists, values, and society / by Leslie Stevenson and Henry Byerly.
        p.   cm.
    Includes bibliographical references and index.
    ISBN 0-8133-2016-X. — ISBN 0-8133-2017-8 (pbk.)
    1. Science—History. 2. Science—Philosophy—History.
3. Science—Social aspects—History.   I. Byerly, Henry.   II. Title.
Q125.S7437   1995
509—dc20
                                                                              94-46510
                                                                                    CIP

Printed and bound in the United States of America

The paper used in this publication meets the requirements of the American National Standard for Permanence of Paper for Printed Library Materials Z39.48-1984.

10   9   8   7   6   5   4   3

Contents

Preface

The initial idea for a book like this arose from Leslie Stevenson designing and teaching a small course-unit entitled "Uses and Misuses of Science" for students in the Faculty of Science in the University of St. Andrews in Scotland. A term's research leave in 1990 (thanks to the University of St. Andrews) enabled him to write a first draft of this book. This time was spent in Macquarrie University in Sydney, Australia, and thanks are due for hospitality and intellectual feedback there. However, with only part of his teaching and research time available for this project, L. S. found it difficult to bring it to completion single-handed.

Meanwhile, Henry Byerly had for some years been teaching a course entitled "Science, Technology, and Values," as well as other courses in the history and philosophy of science, at the University of Arizona at Tucson. Spencer Carr of Westview Press put us in touch with each other, and we quickly found sufficient commonality of approach and complementarity of expertise for a jointly authored project to be attractive. We met for an intensive and stimulating two weeks' collaborative work in St. Andrews in 1993 and again in the somewhat warmer surroundings of Tucson in 1994 (aided by research and travel grants from the School of Philosophical and Anthropological Studies at St. Andrews).

Our collaboration has extended over every section of the book. L. S. drafted about two-thirds of the case studies and philosophical discussion; H. B. added more case studies and historical material. But every sentence has been gone over by both of us (and has been put through several computer systems in the process!). We have thus achieved together something that neither of us would have done alone.

Science shows many faces—it is the work of a great variety of heroic individuals and evolving scientific communities. Science and its works evoke hopes and fears, boundless admiration and deep loathing. In *Science in a Free Society* (1978), Paul Feyerabend asked sardonically, "What's so great about science?" (73). His challenge was to vindicate the supposed special authority of scientific method, an authority that makes its pronouncements more worthy of rational belief than those of other "traditions" such as religions, folklore, and cultural beliefs of any kind. He claimed that no good answer can be found. We are less pessimistic about that (see sections 1.1 and 3.1 of this book), but we do not attempt to settle all the philosophical issues raised by Feyerabend's question. We concentrate instead on the practical side: whether, all things considered, science (both pure and as applied to technology) has been, and will continue to be, a Good Thing for humankind. To that question we offer a fairly positive answer, with qualifications. To echo the request that Oliver Cromwell made of the artist commissioned to paint his portrait, we try to depict science, its practitioners, and its applications as they are, warts and all. But despite the warts, we suggest that there is something "great" about the whole enterprise of science, which has transformed so many of our beliefs and so much of our lives since the seventeenth century.

Although we cannot settle all the philosophical issues that have been raised about science, we do pose the questions and provide some perspective by examining case histories in which the questions arise. We have adopted the policy of making the important issues about science, technology, and human values concrete and vivid by presenting a variety of particular scientists in their historical, social, financial, and political contexts. Support for this concentration on case studies rather than general theorizing can be found from no less a theoretician than the great eighteenth-century philosopher Immanuel Kant. In the "Methodology" section of his *Critique of Practical Reason* (1956), he wrote:

> Those who otherwise find everything which is subtle and minute in theoretical questions dry and vexing [Kant was talking of theory in moral philosophy] soon take part when it is a question of the moral import of a good or bad act that is recounted. . . . I do not know why the educators of youth have not long since made use of this propensity of reason to enter with pleasure upon the most subtle examination of practical questions put to them, and why, after laying the foundation in a purely moral catechism, they have not searched through biographies of ancient and modern times with the purpose of having examples at hand of the duties they lay down, so that, by comparing similar reactions under various circumstances, they could begin to exercise the moral judgment of their pupils in marking the greater or less moral significance of the actions. (153–154)

We have both, for better or for worse, been "educators of youth" for much of our time. As the reader will see, we have indeed been "searching through biographies of ancient and modern times" for vivid examples of the dilemmas, moral and otherwise, that face scientists. Though we have hardly attempted to "lay the foundation in a purely moral catechism" (perhaps we thereby risk Kant's disapproval!), we do have an educative and moral purpose in mind—to stimulate students and other readers to consider critically the value questions that arise in the practice and application of science.

We are philosophers, not historians, and certainly not scientists. But the beginning student can, we hope, learn from this book a certain amount of basic science, history, and politics and make some acquaintance with philosophical issues as they relate to science. Our aim has been to cover as broadly as possible an enormous amount of scientific and historical material at an introductory level; thus, we were faced with difficult choices about what to include and exclude. We have had for the most part to rely on secondary sources, such as nontechnical summaries of scientific theories and their applications, histories of science, and biographies of individual scientists. Some material from Leslie Stevenson's paper "Is Scientific Research Value-Neutral?" in *Inquiry* 32 (1989) has been recycled into Chapter 10.

We are grateful to two reviewers of an earlier version of the manuscript submitted to Westview Press and especially to Spencer Carr, who (unusually for an editor these days) found time to read through the whole manuscript himself. These three gave us very extensive and helpful comments, of which we have tried to make good use. Thanks are also due to Julian Crowe in St. Andrews and Rob Cummins and Ann Hickman in Tucson for help with translating one word-processing system to another.

Leslie Stevenson
Henry Byerly

Introduction

Almost everyone agrees that the development of scientific knowledge has made enormous progress—from understanding the birth of stars and the structure of atoms to explaining the evolution of life on earth and unraveling the biochemistry of the genetic code. But have science and technology improved the human condition, apart from adding to our knowledge? Most people would still probably answer yes, though doubts about whether the overall benefits outweigh the costs seem to have increased in recent years. Surely we now have more wealth and better health than we did 300 years ago—in the developed countries of the West, at least. Some of the applications of science read like a fantastic success story. Consider the "miracles" of modern medicine—antibiotics, the eradication of smallpox, the steady decline of infant mortality, and the lengthening of the average life span; consider also the technologies we now take for granted, from radio and television to CDs, from steam engines to jet planes and rockets that have gone to the moon.

But doubts about the much-vaunted benefits of science and technology have also grown over the years, especially in this century. We have been made all too vividly aware of the enormous destructive power that scientific discoveries have put into military hands—from dynamite and poison gas to missiles loaded with hydrogen bombs sufficient to destroy human civilization many times over, perhaps even to extinguish most life on earth. The more insidious threats from the side effects of science-based industry cause widespread concern: pollution of the environment, radioactive by-products that will remain dangerous for millennia, destruction of the ozone layer, and global warming. Besides the dangers to life and health, worries abound about more subtle ethical problems lurking in the power of scientific knowledge and technology to control people and in the new and difficult choices genetic engineering gives us about human life and reproduction. As the twentieth century draws toward its close, some pessimistic voices can be heard wondering whether the scientific enterprise may be doing more harm than good, all things considered.

The modern era has been called the age of scientific progress and also the age of anxiety. The accelerating expansion of science and technology has provided major sources for progress and hope as well as for worry and dismay. What are the overall results of the revolutions in science over the past several centuries? Science and technology have brought opportunities and threats, benefits and costs. Can we expect continued progress? Are there limits to what science can discover and control in nature? Who and what determine how science is used? Does science merit the awed respect and admiration that it so often receives? Does it deserve the fear and anxiety that it sometimes provokes? In asking these questions and trying to offer some provisional answers to them, we hope to appeal to that native curiosity, present to some degree in all of us, that has inspired scientists in the past: the curiosity to understand, to see connections, and to explain. First, we offer some historical perspectives on the development of science and its role in society (Chapters 1–3). Then we examine a variety of fascinating and instructive episodes in the history of science (Chapters 4–9). We have selected from biographies of scientists and recent scientific journalism vivid examples of the variety of motives and influences involved in scientific research and of the problems and dilemmas raised by scientific discoveries and their applications. We conclude the book with a more philosophical discussion of the relationship of science and values (Chapter 10).

We hope to shed light on the nature and implications of the scientific enterprise and to provide some understanding of what science has been in the past, is at present, and might become in the future. We thus hope to meet criticisms such as Paul Davies has made of the way science is often taught: "Personalities rarely enter. . . . Science is presented as a body of accepted fact, missing out on the anguish, the feuding, the doubt, the exhilaration, and the eccentricity" (1993, 68). Our hope is that students (who sometimes find science boring) will respond to the excitement and the mystery, the competition and the cooperation, the adventure and the frustrations that are actually involved in the pursuit of science.

Our aim throughout this book is to stimulate all our readers—but especially those students of science who will be among the scientists, technologists, or administrators of the future—to think carefully and critically for themselves about the many momentous issues arising from the practice of science.

L. S.
H. B.

1 How Science as We Know It Has Developed

The term *science* is often used in a broad and rather vague sense. To bolster the credibility of claims for a product, advertisements speak of what "science" has shown or what "scientists" say. These words, along with the adjective *scientific,* are now generally used as honorific terms that express intellectual authority and indicate some especially trustworthy way of justifying beliefs. These days we are all expected to accept this, even if we know little or nothing about the theory and practice of science. Occasionally, and perhaps increasingly, voices are heard suggesting that all is not well with orthodox science, that it either ignores or is less than perfectly objective in regards to important and interesting phenomena such as ESP and alternative medicine and that it is subservient to the interests of big business in studying, for example, environmental pollution. And for a long time, some have held the plausible view that science cannot by its nature deal with certain vital human concerns (for example, personal relationships, moral values, and religious faith). But for the most part, scientists, scientific theories, and scientific method retain a special authority in our culture.

Much has been written about the epistemology of science, examining and sometimes questioning the nature of its special intellectual authority (see certain radical philosophers and sociologists of science, such as Feyerabend [1978], Bloor [1976], and Woolgar [1988]). Rather more often, philosophers of science have defended its privileged status of credibility (for example, Popper [1963], Hempel [1966], and Newton-Smith [1981]), even if they admit that articulating the distinctive characteristics of scientific method in an exact but general philosophical account is a difficult task. This book will not attempt to tread this well-trodden ground once more, but we cannot avoid saying something at the outset about what we take science to be, before we go on to explore how it has been pursued and applied by various scientists in different social contexts.

1.1 What Is Science?

What is the aim of science? One short answer is "truth." But what kind of truths do scientists seek, and how and why do they seek them? To

1

get clearer on what science is, let us start with what it is *not*. Science is not technology; it does not consist in the invention of devices, such as those that the twentieth century has brought us: radios, TVs, VCRs, computers, airplanes, antibiotics, lasers, CAT scans, and nuclear weapons. *Technology* (from *technē*, "craft") originally meant systematic knowledge of a craft such as glassblowing or pottery; it is the application of knowledge for practical purposes, to make things and achieve humanly useful results. High tech involves applying scientific knowledge. But in a broad sense, technology includes the use of fire and stone axes and thus antedates science. Unlike science, moreover, technology is common in some form to every human culture that has ever existed. Only since the late nineteenth century has technology become closely tied to science. Nowadays they are so closely associated that people often speak of science-and-technology in one breath.

The central concern of science is "knowing that," the knowledge of truths, as opposed to "knowing how," the knowledge of techniques. Admittedly, these two kinds of knowing are often interconnected, particularly in recent times. Theoretical knowledge is a necessary basis for many contemporary technological applications. And science needs technology; research scientists require experimental know-how and to-day often use elaborate and expensive technologies to test their theories. Nevertheless, we can still distinguish the goal of "pure" science—which aims to *understand* some aspect of the world—from that of "applied" science—which is used to *change* the world in some humanly beneficial way. Of course, some scientists may pursue research of both kinds at different times or may even have both aims in mind in a single project (as we shall see later).

What sort of truths, then, does pure science seek? Science is not merely the collection of observations or data; it is the development and testing of hypotheses, theories, and models that interpret and explain the data. As we have come to recognize it today, science is primarily the attempt to understand the workings of nature by means of general theories. Scientific theories usually involve hypotheses about unobserved entities or processes that may be in principle imperceptible because they are too large, too small, or simply not the sort of things that human sense organs can detect (distant galaxies, the Big Bang origin of the universe, molecules, atoms and subatomic particles, magnetic fields, genes, the evolution of species, and so on). Theories about such things aim to explain what we *can* observe more directly, and they must be confirmed or disconfirmed according to their success in doing so.

Historically, these features can first be clearly recognized in seventeenth-century physics, Isaac Newton's theory of mechanics being the paradigm case. The term *modern science* is often used to refer to science since that time. And it must be emphasized that science is a dynamic process: Theories can always be rejected, modified, or extended. A static collection of theories would be dead science. So, in trying to understand the "game of science," we should focus not so much on particular results—the theories that scientists accept at any given time—but on the way the game is played, the rules that govern it, what the goals are, and what consequences can be expected. We shall examine the scientific approach to questions as a critical attitude toward testing propositions and acquiring knowledge, rather than thinking of science as a collection of facts or established beliefs.

The scientific attitude is one that almost everyone takes at some time—perhaps especially during childhood when we persistently ask "why?" and "how?" But as a steadfast, persevering approach to problems, the critical scientific approach is not common—in fact, it even appears to be not entirely natural as a matter of human psychology. American philosopher Charles Peirce (1839–1914) argued that doubt is central to the scientific mind-set, whereas most people usually try to avoid doubt because it is unsettling, even painful. In "The Fixation of Belief" (1877), he wrote that people have tended to follow "the way of tenacity" (sticking to previously formed beliefs) or "the way of authority" (letting the burden of fixing beliefs fall on someone else). The revolt against doctrines dictated by religious, political, and classical academic authorities was an important component in the rise of modern science.

English philosopher Bertrand Russell (1872–1970) emphasized in his many works on science that it is not *what* a scientist believes that distinguishes him but *why* he believes it. Scientific claims are based (somehow) on evidence as opposed to authority. This critical attitude of science is encapsulated by Karl Popper's phrase "conjectures and refutations." Scientists make hypotheses, which then must meet stringent criticism in the form of observational or experimental testing. Logically, the generality of scientific theories, their goal of stating universal laws, makes them in principle open to falsification by even a single counterinstance. (In practice, disconfirmation of theories is a complex affair.) Likewise, scientific claims must meet standards of quantitative precision that make them readily testable and hence falsifiable. Predicting that an earthquake of a specific magnitude will strike Los Angeles on January 1, 1999, is a stronger test than predicting that an earthquake will strike somewhere in California in the next decade. And since it is not clear what could *possibly* falsify a fortune-teller's vague prediction that next week is a good one for making decisions, such a claim has no scientific status.

A claim does not become scientific simply by virtue of being true. Even if some of the purported astrological correlations between certain positions of the planets and characteristics of persons born under them were to hold up under investigation, astrology would not thereby become science if its central theoretical claims of causal influence between heavenly and human phenomena were not also subjected to critical scrutiny. A mere summary of observed data, however accurate, is not a scientific theory. Conversely, there are many scientific theories of the past, some of which we may still respect as well justified by the evidence available in their time, that we do not now accept. Of course, if we now think a claim is false, we can no longer describe it as knowledge. But the way in which the original hypothesis was formulated and tested might remain an example of good scientific practice (as might be claimed for Ptolemaic astronomy—see 4.1—or the phlogiston theory of combustion—see 4.3).

A special kind of thought process is required to conceive of the very idea of a controlled experiment, which is so basic to scientific research. The following anecdote illustrates this:

> It was a good answer that was made by one who when they showed him hanging in a temple a picture of those who had paid their vows as having escaped shipwreck, and would have him say whether he did not now acknowledge the power of the gods,—"Aye," asked he again, "But where are they painted that were drowned after their vows?" And such is the way of all superstition, whether in astrology, dreams, omens, divine judgments, or the like. (Bacon 1870, vol. 4, 56)

The point is, of course, that a number of successes following a certain course of action does not prove anything unless it is compared with the number of failures. If a certain therapy has apparently worked for many people, we are easily impressed and tend to forget to ask how many people recover from such conditions without any treatment at all or with other sorts of therapy—or indeed, how many people have been made worse by the method in question. Looking at the "control" group seems less natural, and this step is easily neglected. Psychologists have devised subtle tests to show the human tendency to recognize confirmations rather than disconfirmations of hypotheses. Such bias is retained even by trained scientists when they react intuitively, without reflection, and they have to guard against it. This is especially important when the observers themselves may have strong hopes of finding positive results, for example, in investigating extrasensory perception or the claimed success rate of medical treatments (whether "scientific" or "alternative"). This is the rationale for the careful testing of proposed new drugs.

Adherents of doctrines that are suspect sometimes use the label "scientific" in an attempt to appear more credible—for example, scientific creationism or scientific astrology. It is vital to understand how to distinguish science from pseudoscience. Physicist Paul Davies, when asked why he found it "comparatively easy to believe in evaporating black holes and invisible cosmic matter, but not in straightforward things like ghosts and flying saucers that ordinary people see all the time," offered two basic criteria: (1) Scientists, unlike cranks, try to relate their work to existing science, and (2) if a theory differs from accepted views, scientists try to deduce novel predictions by which their hypotheses can be observationally tested (1993, 93–94). Orthodox science is not infallible, and it can never be complete, but any additions or subtractions must ultimately be justified by observation.

Scientists really want to know and to know really. Admittedly, they can't operate without presuppositions, theoretical ideas, and even intuitions. They shape and reshape the questions. But they must have a willingness ultimately to "let nature decide" the answers to their questions. The scientist proposes, but nature disposes. How in detail this works—the logic, methodology, and epistemology of it all—turns out to be very complex and remains controversial, as recent work in the philosophy of science abundantly illustrates. (See Suggestions for Further Reading at the end of this chapter.)

1.2 The Rise of Modern Science

The emergence of modern science appears as something of a miracle when we consider the complex of factors that seem to have been required for it to get started. Unlike technology, science has not been present at every stage of human history or in every cultural tradition. Chinese civilization, for example, has had a rich history in many areas of endeavor. Chinese technology was in some respects more highly developed than that of Europe until the sixteenth century, and it provided some of the keys to European dominance of the rest of the world since then: gunpowder, the magnetic compass, and paper for printing. But science never really developed in China until it was introduced from Western Europe. In a famous letter in which he responded to the question of why science developed in the West and not in China, Albert Einstein remarked: "It is not surprising that science did not arise in other civilizations; it is surprising that it ever arose at all" (quoted in Price 1962, 15).

The beginnings of the scientific tradition can be traced back to ancient Greek philosophers of nature starting with Thales (ca. 600 B.C.). But it is fair to say that science first flourished only with the rise of the new physics in Western Europe in the seventeenth century, beginning with Galileo Galilei and Johannes Kepler and coming into full flower with Newtonian mechanics. This emergence of the scientific tradition is a major component of the modern age, in the historian's sense of the term *modern*. To get some insight into the nature of science and its relation to culture, let us consider some of the historical features of the epoch leading up to the rise of modern science in the seventeenth century. Even a brief sketch of the major causes that have been suggested for the emergence of science shows the complexity of this development. The historical influences are so interwoven that it is difficult to assign priority. Debate continues among historians on whether advances in technology stimulated development of scientific theory or whether it was science that first sparked the invention of new technology; in fact, a complex interplay seems to have existed between science and technology in both directions. We can list the following six major factors, without trying to point to any one of them as "the basic cause" of the rise of modern science.

1. *The resurgence of Greek culture.* The fall of Constantinople (now known as Istanbul) to the Turks in 1453 is often cited as a turning point of the Renaissance. A flood of refugee scholars into Italy brought to Europe many works of philosophy, mathematics, and astronomy that had been written by the ancient Greeks nearly two millennia earlier. Knowledge of Euclid, Archimedes, and hitherto unknown works of Aristotle and Plato stimulated a rebirth of "natural philosophy" (as scientific inquiry was then called). Arabic scholars, especially those working on the border of the Christian and Moslem worlds in Spain, were also a vital source for the recovery of the works of the ancient world by the West.

2. *The invention of the printing press.* Johannes Gutenberg's Latin edition of the Bible appeared in 1455. Printing spread rapidly in the late fifteenth century, making possible the communication required for rapid progress in science. The printer was a prototype of the early capitalist; printing was one of the first examples of mass production, and an effective print shop required considerable capital expenditure. A necessary condition for the early technology of printing was a phonetic alphabet, which may help explain why science did not develop first in China. Cultural critic Marshall McLuhan (1962) even suggested that a phonetic alphabet was the ultimate source of "mechanistic" science.

3. *The rise of capitalism.* One connection of science with capitalism was the development of mathematics to serve the needs of commerce, such as the calculation needed in accounting. The arithmetical notation we use today—fractions, decimals, and the equal sign—were introduced in the sixteenth century. Capitalist enterprise gave rise to a new middle class, which included most of those who were receptive to the new scientific developments—tradesmen, lawyers, and doctors. By contributing to the secularization of society, with a focus on success in this world rather than salvation in the next, capitalism encouraged attempts to understand and control the natural environment.

4. *The discovery of the Americas.* An increase in world trade, especially after the voyages to the New World begun by Christopher Columbus in 1492, stimulated inquiry by the discovery of new plants, animals, and cultures. European foreign trade tripled in the seventeenth century, and the increase in oceanic navigation gave rise to the need for more accurate clocks and astronomical calculations.

5. *Monotheism.* The concept of one Creator of the universe (an idea common to Judaism, Christianity, and Islam) might be thought to support the scientific project of using human reason to inquire into the divine design of nature. But this support had to wait for the right historical conditions. Scientific inquiry does not appear to have been encouraged in Judaism or in pre-Reformation Christianity. To some religious casts of mind, it was presumptuous, even idolatrous, to use human reason to seek knowledge of God and his works. Rather, as previously noted, it was Islamic scholarship that maintained and extended the ancient Greek tradition, thus giving the initial impetus to the early development of science, mathematics, and medicine. Some of the great names were Avicenna (980–1037) and Averrhoës (1126–1198). Our mathematical terms *algebra* and *algorithm* come from Arabic. The early Islamic religious tradition had a commitment to knowledge transmitted through written language, which inspired renewed scholarship in the West in the late Middle Ages.

6. *The Reformation.* The Protestant break from Roman Catholicism in the sixteenth century ended the latter's monolithic authority and encouraged individual inquiry. There was then a turn from "revealed" theology, mediated by the church, to "natural" theology, which praised the glory of God by seeking evidences of the Creator's design in nature. The Reformation also indirectly contributed to the development of modern science by supporting the formation of nation states, which fostered

secularization. And the Protestant ethic has been associated with both the emergence of capitalism and the cultivation of natural science (Merton 1957).

Other contributing elements might be cited. One important technological advance was the telescope, invented around 1600 in Holland. Galileo heard of this new tool and built one for himself. This allowed him to "bring the heavens down to earth," as historian and scientist J. D. Bernal (1965, 292) put it: "In the first few nights of observation of the heavens [with his telescope] Galileo saw enough to shatter the whole of the Aristotelian picture"—craters on the moon; sunspots; phases of Venus; and as a visible model of the heliocentric Copernican system, the orbiting of the moons of Jupiter.

Science was not at first sharply distinguished from what we would now call pseudoscience, and the same individual could be active in both. Kepler (see 4.1) was employed as an official astrologer, although his astrological activity was separated from his work in physics and astronomy. Likewise, Newton (see 4.1) spent much of his time studying alchemy, which had as a major goal transmuting base elements into gold. He has been called the last of the great magicians, those who try to master occult powers. He wrote millions of words on his religious and metaphysical speculations, but virtually all his published works explored physics as we now recognize it.

Nor was science yet clearly distinct from philosophy. René Descartes (1596–1650) was one contributor to the scientific revolution who is equally famous as a father of modern philosophy. He saw his system as a tree with metaphysics as its trunk and the sciences as its branches. Newton's great *Principia* contains methodological discussions that were vital to his whole project of explaining nature by mathematical principles but that would today be classified as philosophy of science rather than as part of physics.

In those early days of science, research was pursued by those who had the interest, the leisure, and the means to do it. There were hardly any professional positions available. Newton was fortunate to be elected in his twenties to a lifetime professorship in mathematics (and a well-paid college fellowship) at Cambridge University in England. But his predecessor Galileo had to scheme to find favor with regional governments and aristocratic private patrons in early seventeenth-century Italy—with catastrophic results that we shall examine in 7.1. Underscoring the limited career opportunities in science at the time is the fact that the term *scientist* had not yet come into use—*natural philosopher* was the nearest equivalent.

As we shall see in 2.1, Francis Bacon was a prophet of both scientific research and its application to human benefit who tried unsuccessfully to persuade the powers-that-were to institutionalize his vision. But after his death, his idea of a college of experimental philosophy inspired a group of English intellectuals to form the "Invisible College." Mathematician John Wallis (1616–1703) described this "college" as meetings of "diverse worthy persons, inquisitive into natural philosophy and other parts of human learning, and particularly of what has been called the New Philosophy or Experimental Philosophy" (Hall 1962, 193). This group developed into the famous Royal Society. In 1662, King Charles II gave a charter to the Royal Society of London for Improving Natural Knowledge, and it became the first of the scientific societies or academies that sprang up in Europe in the late seventeenth century (the French Academie soon followed). The Royal Society encouraged scientific investigation in diverse areas ranging from husbandry and botany to chemistry and astronomy and fought the evils of superstition, such as the persecution of supposed witches.

The needs of navigation were important in encouraging further developments of the New Science. The first nationally financed scientific institution, the Observatoire Royal at Paris, was established for this purpose in 1672, and three years later the Royal Observatory at Greenwich was completed. A central problem was the determination of longitude. By comparing local solar time with the time on a clock that had been set at Greenwich mean time at a known longitude, local longitude can be calculated. Before the advent of radio communication, however, there were only two methods for determining Greenwich mean time: to have at hand an accurate clock or to observe certain regular movements of the heavenly bodies as a substitute, in effect a clock in the sky (for example, the movements of the moon—or as Galileo proposed, the satellites of Jupiter). Many investigations of the early Royal Society were concerned with determining longitude. The problem was actually solved technologically rather than scientifically by the invention in the eighteenth century of the chronometer, a device that kept accurate time at sea in all conditions.

With the important exception of navigation, most scientific research in its early flowering in the seventeenth century was not directly aimed at being useful. Intellectual curiosity, rather than benefit to humanity, profit, or power, was the main motivating factor. By 1700, experimenting and gathering facts for the advancement of knowledge was quite the fashion among intellectuals in England.

1.3 The Professionalization of Science

Newton obviously counts as a scientist in anyone's book, but in 1687 his new science of mechanics was published in *Mathematical Principles of Natural Philosophy* (translated from the Latin title). The English word *science* derives from the Latin word *scientia*, which simply means "knowledge." Today, the notion of science is more restrictive; it does not cover historical knowledge, for example, and pure mathematics is often not counted as a science proper because it does not deal directly with the material world. As we shall see in Chapter 10, it is a matter of heated philosophical debate whether all genuine human knowledge amounts to some kind of science or whether there are kinds of knowledge not accessible by scientific methods.

The first recorded use of the term *scientist* was by British philosopher-scientist William Whewell (1794–1866), after what we now call science had been flourishing for some two hundred years. Whewell suggested referring to practitioners of science as scientists, on the analogy of art and artists, at a meeting of the British Association for the Advancement of Science in 1833. The very concept of a scientist has evolved historically and may still be changing in our own time. At the beginnings of modern science in England, we could have looked at a list of the fellows of the Royal Society to see who then counted as prominent natural philosophers. However, during the seventeenth and eighteenth centuries, the scientific academies enlisted many who were not active practitioners of science. Early members of the Royal Society had a remarkable diversity of social and educational background—they were bishops, architects, poets, and gentry—but given the class divisions of that period, it was difficult for people to make their mark in science, however able they might have been, without patronage.

Only in the nineteenth century did science begin to emerge as a standardly recognized profession, primarily in the universities. Even there, the sciences had at first to struggle for acceptance as academic subjects alongside the traditional liberal arts that had been taught since medieval times, but following the lead of the reorganized universities in Germany, this trickle of change became an avalanche. One indication of this professionalization of science is that the criteria for membership of the Royal Society were revised around 1850 so that only people with recognized scientific stature would be admitted.

Thus, by the late nineteenth century, many of those recognized as scientists had professional positions as such—in universities or as officials of national scientific academies or museums. But there were still "gentleman" scientists—and they were, of course, virtually all men—who were not paid for doing their work, who were amateurs in the sense that they did it for love rather than money, although their contributions may have been highly professional in the sense of being very competent. Charles Darwin is a prime example; he had sufficient private means to live as a country gentleman, but he devoted himself full-time to his research, keeping in constant touch with other leading researchers in biology and geology (see 7.1).

Because of the burgeoning technicality of the various sciences and the constant development of knowledge in them, it became increasingly difficult for someone without full-time training and employment in science to achieve scientific results and get them recognized. Today, it is a virtual necessity to have a Ph.D. and a paid position in some scientific discipline to be a recognized scientist. Because of the massive accumulation of knowledge, students must now spend many years absorbing what they need to know about what has already been achieved in any given branch of science. (In his highly influential work *The Structure of Scientific Revolutions*, Thomas Kuhn [1962] rightly rejects simplistic accumulative models of the development of science—but the quantity of scientific knowledge has certainly increased, however devious the process of theory revision.)

Few people without a Ph.D. try to get their work published in recognized scientific journals, and fewer still succeed. The case of Albert Einstein, a then-unknown employee of the Swiss Patent Office who startled the world of physics in 1905 by having three papers accepted by *Annalen der Physik*, was exceptional even then and would be far more unlikely now. It is more difficult today even for brilliant students to advance as quickly to leading positions in science as was once common; many of the great scientists of the past, such as Kepler, Newton, James Clerk Maxwell, William Thomson (Lord Kelvin), and Josiah Gibbs, became full professors by the age of twenty-six.

Identifying the recognized scientists in a given discipline could be roughly accomplished by looking through that discipline's journals, although this method has become increasingly problematic as journals proliferate and the lists of contributors in large research projects lengthen. Among the latter, a relatively small core of individuals do the genuinely innovative work. According to one rough estimate, 10 percent of those working in a field contribute more than half the scientific publications, and probably only a small proportion of the publications represents significant new achievements.

1.4 The Industrialization of Science

The new era of "industrialized science" can be traced back at least to the emergence of the science-based chemical industry in Germany in the late nineteenth and early twentieth centuries. This process, in which Fritz Haber was a typical figure (see 6.2), was hastened by the establishment of the Kaiser Wilhelm Gesellschaft for scientific research, which was funded by German banks and business concerns. This organization was also harnessed to the German effort in World War I. In the United States, the industrialization of science was marked by the national effort on a huge economic scale to produce the first atomic bomb in time for use in World War II (see 9.1). And this trend has become ever more obvious since then in many areas of science.

Much scientific research nowadays involves high technology—electron microscopes, ultracentrifuges, particle accelerators and colliders, telescopes mounted on satellites, and the like—which makes scientific progress in any area heavily dependent on the proper functioning of the relevant technology, and hence on specialist technicians as well. All this costs a lot of money, much more than the average university can find, so funding must now often be sought at the national or even international level. But governments and industries do not hand out big money to fund "Big Science" without expecting something in return, whether in the form of profit, medical benefit, national security, or whatever. Science funding thus becomes a matter of national policy. Of necessity, this sort of science is subject to industrialization either quite literally, in that the research project is undertaken by private corporations for their own ends, or more metaphorically, in that even if publicly funded, it has to be organized and planned on an industrial scale, with large teams of people working with specialized apparatus.

So even in "pure" areas of science, results are often a product of large research teams, with each member contributing a piece to solving the puzzle. Scientific papers are now typically published in the name of many individuals, including all those in the research team who have helped achieve the result. The work of particular individuals still occasionally stands out, as in the case of some Nobel Prize winners, but the interdependence of many contemporary contributions contrasts with the research process of earlier times. By its nature, scientific knowledge is dependent on previous discoveries that can in turn be traced back to the work of the early heroes of science. The increasingly rapid and complex growth of scientific and technological activity has resulted in an enterprise that today exhibits a very complex web of intellectual, technical, economic, and political relationships. We shall be examining case studies to illustrate this in Chapters 4 through 9.

Suggestions for Further Reading

On science and society:

Conant, J. B. 1951. *On Understanding Science.*
Feyerabend, P. K. 1978. *Science in a Free Society.*
Merton, R. K. 1957. *Social Theory and Social Structure.*
Price, D. de Solla. 1963. *Little Science, Big Science.*
Ravetz, J. 1971. *Scientific Knowledge and Its Social Problems.*

On scientific method:

Chalmers, A. F. 1982. *What Is This Thing Called Science?*
Hempel, C. G. 1966. *Philosophy of Natural Science.*
Newton-Smith, W. 1981. *The Rationality of Science.*
Popper, K. R. 1963. *Conjectures and Refutations.*
Salmon, W. C. 1975. *The Foundations of Scientific Inference.*

On the history of science:

Bernal, J. D. 1965. *Science in History.*
Burke, J. 1978. *Connections.*
Butterfield, H. 1957. *The Origins of Modern Science.*
Feuer, L. S. 1992. *The Scientific Intellectual: The Psychological & Sociological Origins of Modern Science.*
Hall, A. R. 1962. *The Scientific Revolution, 1500–1800.*
Price, D. de Solla. 1962. *Science Since Babylon.*
Sarton, G. 1936. *The Study of the History of Science.*

2 *Images of Science*

Many different views have been taken, and still are taken, of the scientific enterprise—its costs and benefits, its relations to the rest of human society and culture, and its overall value to humanity. In this chapter we will survey some of the most influential images of science. Critical evaluation of these various views is, for the most part, reserved until the final chapter.

2.1 The Baconian Vision: Science as Bountiful

People of the ancient world had little expectation of progress as we think of it. They tended to look backward to the so-called Golden Age and preferred cyclic views of history. There was even a tradition, championed by Plato, of viewing change as degeneration. The only hint of belief in progress came from thinkers such as Archimedes, who envisioned successors adding to the knowledge that he had attained. The Judeo-Christian tradition also tended to expect corruption rather than improvement, given its doctrine of Adam's Fall from a primal Paradise. For Christianity, there could be only one true kind of progress—toward salvation; God, not human endeavor, was believed to shape the course of history. But the modern age (which began in the seventeenth century) is marked by a belief in human progress. In the nineteenth century, faith in science reached a high point, as expressed in the declaration of English mathematician and writer on science W. K. Clifford that "scientific thought is not an accompaniment of human progress, but human progress itself" (1885).

Francis Bacon (1561–1621) was one of the first and most influential thinkers to proclaim this faith in progress through the application of the new science, the new approach to knowledge of nature. He was a lawyer and statesman who rose to the position of lord chancellor of England under King James I, but he was discharged in disgrace at the age of sixty after being convicted of taking bribes. Bacon was not much of a scientist himself, though he is said to have died serving experimental science after coming down with bronchitis caused by stuffing snow in a chicken to test its preservative power! And his precepts for scientific method were not adopted even by those he helped inspire, such as Newton. Bacon underappreciated the role of mathematics in science, which perhaps accounts for his rejecting the astronomical theories of Nicolaus Copernicus and Kepler. Yet despite this, he can be seen as the herald of the modern age's scientific spirit. He himself said, "I rang the bell which brought the wits together" (quoted in Edwards 1967, 239).

Bacon called his plan for a new science the "great instauration," that is, the renewal of knowledge that would restore the human race to mastery over nature. But what he proposed was something quite new: "a total reconstruction of the sciences, arts, and all human knowledge raised upon the proper foundation" (*Works*, vol. 4, 8). Traditional philosophy alone would not serve this purpose, he argued, because "all the disputations of the learned never brought to light one effect of nature before unknown" (vol. 1, 123). (Descartes, his French contemporary who developed a much more mathematical program for scientific method but with rather less emphasis on practical applications, made a similar complaint: "There is not a single matter within [philosophy's] sphere which is not still in dispute" [1955, 86].)

The mechanical arts, by contrast, had made continual (though slow) progress through trial and error. In the Middle Ages, a number of creative advances in technology led to such innovations as improved plows, windmills, spinning wheels, and mechanical clocks (first invented in China). What was needed, Bacon saw, was to join "head and hand," to relate the systematic thought and speculative theory of the scholars to the observation and experience of the artisans and thus achieve a balance between "light" and "fruit." "The unassisted hand and the understanding left to itself possess but little power," he affirmed, yet "knowledge and human power are synonymous" (1870, vol. 4, 47).

According to Bacon, science can and should "transform the condition of human life by gaining and applying knowledge of nature" that would grant us "new discoveries and powers" (vol. 4, 79). Whoever could "succeed in kindling a light in nature" and thereby "illuminate all the border regions on the circle of our present knowledge" would be "the propagator of man's empire over the universe, the champion of liberty, the conqueror and subduer of necessities" (quoted in Farrington 1969, 40). He wrote of his hope for the future in a story about travellers who discover a utopian community on the mythical island of New Atlantis. They are taken to visit the House of Salomon, which is in effect a well-organized scientific research center, whose purpose is described in the following oft-quoted passage, typical of Bacon's eloquence and ambition: "The end of our foundation is the knowledge of causes, and secret motions of things; and the enlarging of the bounds of human empire to the effecting of all things possible" (vol. 3, 156). This vividly expresses the theme of the Baconian vision of scientific progress: Knowledge is power—it gives people the ability to produce all manner of good things for the common benefit. One can't say he didn't aim high!

Bacon's optimism contrasts sharply with the gloomy views of humankind and nature that were common at the time. The poet John Donne (a contemporary of Bacon) expressed the traditional Christian view that the world and its inhabitants had been decaying since the Fall, awaiting the final judgment day. But Bacon hoped to reverse the direction of history since the Fall and to encourage us to work for paradise on earth. He heralded a new attitude toward history, an attitude that held the book of Ecclesiastes in the Old Testament to be quite wrong—there shall indeed be new things under the sun. The idea of scientific knowledge as liberating the human spirit gave a new twist to the biblical promise: "You shall know the truth, and the truth will set you free" (John 8:32 NAS). Like the explorers of distant shores, the scientist was an adventurer in a quest to conquer new frontiers of knowledge. The frontispiece of Bacon's *Novum Organum* (1620)—"the new instrument of scientific method"—represented this spirit of adventure. It shows the Straits of Gibraltar with the motto *plus ultra* (there is more beyond), replacing the pre-Columbian conception *non plus ultra* (there is nothing beyond).

As a spokesman for the emerging scientific view, Bacon was remarkably effective. His eloquent rhetoric, often quoted to this day, had a powerful influence on the generations that followed. He saw a new world coming into being. Indeed, it is striking just how much of his vision has since come to pass. But in addition to all this optimistic eloquence, which so many apostles of scientific progress have loved to quote, there is a darker side to Bacon's language that feminists and ecologists have recently brought to our notice (see especially Merchant 1980, ch. 7). Bacon represented science and technology as the interrogation and exploitation of nature (represented as female), and his language was often remarkably violent and sexual, going so far as to compare nature under scientific "inquisition" to a suspect interrogated under torture: "For like as a man's disposition is never well known or proved till he be crossed . . . so nature exhibits herself more clearly under the trials and vexations of art [that is, mechanical devices] than when left to herself" (298). (Bacon was writing at a time when his sovereign, James I, whom he hoped to influence, was encouraging greater severity in the trials of witches.) His favorite theme of human dominion over nature was often given sexually suggestive expression, as in this example: "By art and the hand of man . . . nature can be forced out of her natural state and squeezed and molded" (29). Sometimes, however, he made a more peaceful, domestic application of his favorite metaphor of scientific masculinity: "What I purpose is to unite you with things themselves in a chaste, holy, and legal wedlock; and from this association you will secure an increase beyond all the hopes and prayers of ordinary marriages, to wit, a blessed race of Heroes or Supermen who will overcome the immeasurable helplessness and poverty of the human race" (quoted in Farrington 1966, 72). But it has to be admitted that Bacon's more violent language does all too clearly prefigure the "rape of nature"—the ecological devastation that has become obvious in this century (see 9.3).

For all his faults, Bacon foresaw that great benefits could be gained from systematic scientific research and the application of scientific knowledge in the exploitation of natural resources. He gave early expression to the hope that if only human circumstances could be controlled, humanity itself could be perfected. In the eighteenth century, often called the Age of Enlightenment, the Baconian vision dominated Western thought. French economist Anne-Robert-Jacques Turgot (1727–1781) proclaimed that all humanity "marches always, though slowly, towards greater perfection" (quoted in Wiener 1973, 639). The spirit of the new science was prominent among the leaders of the American Revolution. Thomas Jefferson (1743–1846), a man of many talents—author of the Declaration of Independence and an inventor who was knowledgeable of the science of his time—had portraits of the individuals he held to be "unquestionably the three greatest men the world had ever produced": Bacon, Newton, and John Locke, who were for Jefferson, respectively, the chief apostles of reason, science, and liberty (Wernick 1993, 82). Benjamin Franklin (1706–1790), whose research in electricity earned him a prominent place in the science of his time, expressed his faith in scientific progress in a letter to English chemist Joseph Priestley in 1780:

> The rapid progress true science now makes occasions my regretting sometimes that I was born too soon. It is impossible to imagine the height to which may be carried, in a thousand years, the power of man over matter . . . all diseases may by sure means be prevented or cured, not excepting even that of old age. . . . O that moral science were in as fair a way of improvement, that men would cease to be wolves to one another, and that human beings would at length learn what they improperly call humanity.

Bacon's optimistic vision lives on today in statements like "The great progress of our civilization is directly related to the centrality of science in our culture" and in the learned judgment of the historian of science George Sarton, who made an even stronger claim: "Scientific activity is the only one which is obviously and undoubtedly cumulative and progressive" (1962, 10). In 1969, inventor, writer, and lecturer R. Buckminster Fuller (1895–1983) rhapsodized in a Baconian vein on the great prospects that science still offers for the future: "Humanity's mastery of vast, inanimate, inexhaustible energy sources and the accelerated doing more with less of sea, air, and space technology has proven Malthus [see 2.2] to be wrong. Comprehensive physical and economic success for humanity may now be accomplished in one-fourth of a century" (340).

In some respects, twentieth-century science as practiced in today's universities and government and industrial laboratories fulfills Bacon's vision much better than the science that began to flourish shortly after his death. But the utopia that he hoped for, in which nature would be harnessed for the benefit of all humanity, has not so clearly come to pass.

2.2 Frankensteinian Nightmares: Science as Demonic

Not everyone believes that science will be our salvation or even that it is a force for human benefit overall. Whether justified or not, a widespread loss of faith in progress through science and technology has occurred in recent years. If this is the scientific age, what are its fruits? Is there really less misery today, all in all? Surveying the human condition in the twentieth century, we find much to shatter any dreams of utopian progress. On a large scale, there have been two global wars. World War I shook the complacent belief of the late nineteenth century that lasting peace, prosperity, and scientific progress were in reach. And then under Nazism during World War II, Germany, one of the leading nations in science, degenerated into vulgar militarism and the horrors of the concentration camps. Meanwhile in the Soviet Union, followers of the supposedly "scientific" theory of Marxism starved and enslaved millions of their fellow citizens.

At the height of the Cold War, with the nuclear arms race between the United States and the Soviet Union, there were real fears that civilization—and perhaps the human race itself—might be wiped out. Forget about a bright new future, people thought—there might be no future at all! With the breakup of the Soviet Union, the threat of nuclear war has now shifted to more localized conflicts, but the potential of nuclear and chemical and biological weapons proliferating around the world still casts a pall over the future. It is sobering to consider this question: If the construction of hydrogen bombs should ever result in the destruction of civilization or of all human life, would the whole development of modern science have been worthwhile? American historian Henry

Adams, writing in 1862, was remarkably prescient in foreseeing the potential disaster: "Man has mounted science, and is now run away with. I firmly believe that before many centuries more, science will be the master of man. . . . Some day science may have the existence of mankind in its power, and the human race commit suicide by blowing up the world" (quoted in Marx 1964, 350).

Even if the threat of nuclear holocaust has receded (for the time being), there remain problems enough: poverty and starvation in many parts of the world and the senseless violence, individual and collective, with which we are daily bombarded on TV news. A widespread pessimism exists that in its very success in spurring growth—of production, resource consumption, and population—science has also sown the seeds of disaster. Obviously, at *some* point there must be an end to growth. English economist Thomas Malthus (1766–1834) argued that nature limits the potentially geometrical growth of populations through starvation, since the food supply can only grow arithmetically (linearly). He was right about the potential growth of human population; in fact, since his day it has grown vastly more than he believed possible—and it has grown most quickly in the poorest regions. But he reckoned without the application of technology to agriculture, which has (so far) helped increase the food supply to keep approximate pace with population growth. But the basic conclusion is unavoidable: Exponential growth of population must eventually cease, by one means or another.

Foreboding about the consequences of seeking knowledge of nature's secrets appears in various ancient myths. The Greek legend of Prometheus ("Forethinker") tells of a Titan who steals fire, the divine spark from heaven, bringing it down to earth for the benefit of humans. This angers Zeus, father of the gods, who decides to punish the human race by sending down evils in the guise of a woman. He has a maiden fashioned out of clay: Pandora ("All gifts"). She is sent to Prometheus's brother, Epimetheus ("Afterthinker"), who rashly accepts Pandora against Prometheus's warning. Pandora removes the cover of a jar containing a host of evils, letting all of them loose on the world—excepting only Hope. It is not clear whether hope also was considered an evil—it might have been taken to be ultimately deceptive in a fatalistic view of the world. For acting against his wishes, Zeus punishes Prometheus by binding him to a mountain where every day an eagle tears at his liver, which renews itself by night for further torture.

There are many themes to this legend. Prometheus is often taken as a symbolic hero of both the arts and the sciences, the one who brought humanity out of its state of primitive ignorance. He represents the spirit of freedom and knowledge opposing tyranny—but with the threat that those who presume to partake of divine knowledge (who "play God") will be punished by the dangerous consequences of new technologies. For obvious reasons, then, this story continues to strike a chord in us, as typified by this statement: "Never before have the Prometheans [modern scientists] been so daring. Never before have the Epimetheans been so rash and never has Pandora's jar been so crammed with menace" (de Ropp 1972, ii). Crudely, the moral of the story might be read as "Don't play with fire!"

The account of the Fall in the Bible has also been read as a warning that forbidden knowledge is dangerous. God had warned Adam that "of the tree of the knowledge of good and evil you shall not eat, for in the day that you eat of it you shall die" (Genesis 2:17 RSV). Eve, however, was tempted by the serpent, who said, "You will not die. For God knows that when you eat of it your eyes will be opened, and you will be like God, knowing good and evil" (3:4–5). So Eve ate the forbidden fruit and gave some to Adam. For their disobedience, God proclaimed, "Cursed is the ground because of you. . . . In the sweat of your brow you shall eat bread till you return to the ground, for out of it you were taken; you are dust, and to dust you shall return" (3:17–19). Man and woman, having gained knowledge but lost their innocence, were ejected from the paradise of the Garden of Eden.

The famous Gothic horror story *Frankenstein, or the Modern Prometheus* by Mary Wollstonecraft Shelley (1797–1851) is a classic articulation of fear about where scientific ingenuity—not just misapplications of scientific knowledge, but the process of scientific research itself—might lead us. Written in 1816 as a contribution to a ghost story competition among the Shelleys and Lord Byron, the book has been popularized in several film versions. Shelley presented an image of the scientist as determined to gain hidden and potentially dangerous knowledge. Victor Frankenstein is so obsessed with his research project that he is prepared to risk his own safety and even that of his whole society to test to the utmost his ability to understand and control nature. He remarks at the outset of his study of the nature of life: "So much has been done . . . more, far more, will I achieve: treading in the steps already marked, I will pioneer a new way, explore unknown powers, and unfold to the world the deepest mysteries of creation" (1994, 40).

Victor Frankenstein is "delighted in investigating the causes of natural phenomena" (27). Driven by curiosity "to learn the hidden laws of nature," he views the world as "a secret which [he] desired to divine" (27). He studies alchemy, seeking the philosopher's stone (which was supposed to have a magic power to transmute base metals into gold.) Failing in this research, he turns to the science of Newton. Echoing Bacon,

Frankenstein sees in science the power to "penetrate into the recesses of nature, and show how she works in her hiding places" (40). For two years he labors to construct a gigantic living man, getting his materials by nighttime visits to graveyards. Though he finds this occupation loathsome, Frankenstein says of himself:

> With unrelaxed and breathless eagerness, I pursued nature to her hiding places. . . . a resistless, and almost frantic, impulse urged me forward; I seemed to have lost all soul or sensation but for this one pursuit. . . . I wished, as it were, to procrastinate all that related to my feelings of affection until the great object, which swallowed up every habit of my nature, should be completed. (48–49)

When he has at last "discovered the causes of generation and life" and made himself "capable of bestowing animation upon lifeless matter" (45–46), Frankenstein creates a living creature that has human feelings and reason but is grotesque in appearance, having been assembled out of incongruous parts. Horrified by the monster he has made, Frankenstein flees from it. The monster, unloved and resentful, gets out of control and ends up murdering the people most dear to his creator. When Frankenstein threatens to kill it, the creature responds, "I was benevolent and good; misery had made me a fiend. Make me happy, and I shall again be virtuous" (101). Frankenstein then decides to take responsibility for caring for his creation and thinks of creating a wife for him, but just before breathing life into the female creature, he is struck with the fear that he might start a race of monsters capable of destroying humankind. He tells the monster, "Never will I create another like yourself, equal in deformity and wickedness." The monster replies: "Slave, I before reasoned with you, but you have proved yourself unworthy of my condescension. Remember that I have power; you believe yourself miserable, but I can make you so wretched that the light of day will be hateful to you. You are my creator, but I am your master;—obey!" (178–179). After all these frightful events, on his deathbed Frankenstein is still torn two ways, and he says: "Seek happiness in tranquillity and avoid ambition, even if it be only the apparently innocent one of distinguishing yourself in science and discoveries. Yet why do I say this? I have myself been blasted in these hopes, yet another may succeed" (236).

Nowadays, many people find such mythic and romantic forebodings amply justified by events in the present century. When we think of the enormous amounts of scientific expertise and resources that have been devoted to perfecting nuclear bombs, the continuing research into chemical and biological weapons, and the possibilities for genetic engineering to produce new kinds of plants and animals and perhaps to manipulate human genes, then the Frankensteinian image of obsessive and dangerous scientific research may seem more appropriate than that of white-coated purity and beneficence.

Some have argued that behind much of the so-called progress of science and technology there lurks a distinctively masculine, aggressive urge to dominate, penetrate, and make use of the innermost secrets of passive, feminine nature. (We noted such a tendency in Bacon's language in the previous section.) Brian Easlea has interpreted the production of the atomic and hydrogen bombs in terms of such macho impulses: "The principal driving force of the nuclear arms race is not the brute fact of scarce material resources, important though it is, but masculine motivation—in essence, the compulsive desire to lord it over other people and non-human nature, and then manfully to confront a dangerous world" (1983, 165). Easlea finds Mary Shelley's story of Frankenstein full of insights on the obsession with technical mastery that he thinks is typical of so much contemporary scientific and technological activity. The danger is that such an impersonal orientation can involve a repression of ordinary human affections and concerns, which tends to result in disaster.

For somewhat similar reasons, American environmental activist Jeremy Rifkin has argued that our two most powerful technologies—nuclear weapons and genetic engineering—should not be used any further. He finds them objectionable in principle because they encourage our tendencies to control and manipulate human beings in the name of "security" or "efficiency" or even "health," while fundamentally going against our humanity (recall Benjamin Franklin's comment in 2.1 on what people "improperly call humanity"):

> Setting our minds free from the old way of thinking about technology will not be easy. It will require an acknowledgement on our part that some technologies ought not to be used under any circumstances, because what they do threatens either the sacred quality of life or the survivability of life. Certain technologies are so inherently powerful that, in the mere act of using them, we do damage to ourselves and our environment. (1985, 93)

There is no admissible place in Rifkin's view for nuclear weapons or genetic engineering in a world that defines itself as a single family, living in a single community, inhabiting a common ecosystem, and enclosed in a single globe.

Rifkin is deeply pessimistic in seeing no good use for certain technologies. Moreover, he is worried not just about certain applications of science but about the way that science itself is practiced. Nevertheless, he has expressed the hope that we can change our attitude toward science and its uses. He believes that it is possible for us to think in a radically different way, no longer seeking knowledge to obtain power by manipulating our environment in the Baconian tradition but rather by empathizing with it. He seems to have in mind sciences like ethology and ecology, which study the working of whole environmental systems in nature, rather than taking life and matter apart to discover what makes their tiny constituents tick (see 9.3). He has also recommended the adoption of technologies like wind and water power that make use of natural events without radically disturbing them (but such projects would have to be kept small in scale to satisfy this condition). Rifkin has been very successful in using the courts to block research and production in biotechnology, which has caused *Time* magazine to label him as "the nation's foremost opponent of environmental neglect and genetic engineering" as well as "the most hated man in science" (Thompson 1989, 102). He represents one of the loudest contemporary voices of scientific pessimism.

Whatever the reasons for suspicious attitudes toward human motives in science—whether interpreted as overaggressive masculinity, overweening pride, or lust for fame or power or money—mistrust certainly abounds these days. Recalling the biblical story of the garden of Eden, some people may wonder if there are truths that it is better for us not to know. Clearly, we cannot now rest content with naive Baconian optimism; we must find ways of wrestling with the dark possibilities of science and technology.

2.3 Science as Undercutting All Human Values

In addition to worries about destructive or malevolent tendencies of modern science and technology, a different sort of concern has arisen about the overall philosophy or worldview that scientific understanding seems to involve. Many people see it as a threat to all human values. Do we have to accept that we are mere specks of dust in a vast universe that is totally indifferent to our concerns, as science would seem to indicate?

An early expression of dismay with this new image of nature is John Donne's poem about what was then called the New Philosophy, that is, the seventeenth-century approach to scientific knowledge of the world:

And New Philosophy calls all in doubt
The Element of fire is quite put out;
The sun is lost, and th'earth, and no mans wit
Can well direct him where to looke for it.
(An Anatomy of the World, First Anniversary, 1611)

Donne was reacting to Galileo's report of seeing four moons orbiting Jupiter through his telescope, an observation that supported the Copernican cosmology in showing that not everything revolves around the earth. The concern voiced by Donne was that humanity thereby lost its traditional place at the center of the cosmos and was left, as it were, drifting nowhere in space. French mathematician and religious thinker Blaise Pascal (1623–1662), himself active in his youth in scientific investigation, was also troubled by the picture presented in the new cosmology, remarking in his famous *Pensées* ("Thoughts") that "the eternal silence of these infinite spaces terrifies me" (iii.206). The human race seemed no longer to be at the center of the cosmic drama, literally or metaphorically—for the new physics of Galileo and Newton describes the universe as operating according to mechanical laws that take no account of human aspirations or purposes. A major source of hostility to science and the scientific attitude has been this fear that in the very process of trying to understand nature's mysteries, science divorces us from nature.

There were some poets in the so-called Age of Enlightenment, such as John Dryden and Alexander Pope, who welcomed the new physics and its vision of cosmic harmony. (Dryden was himself a member of the Royal Society.) But romantic reaction set in strongly during the late eighteenth century, a counterenlightenment as it has been called. The romantic thinkers were repelled by what they saw as the disenchantment of nature by the dissection and rational analysis of scientific method. English poet, engraver, and mystic William Blake (1757–1827) was horrified by the ugly effects of the early industrial revolution in England, and he associated these with Newtonian mechanics. This attitude is exemplified in this extract from his poem "Jerusalem," written in 1827:

> *I turn my eyes to the Schools and Universities of Europe*
> *And there behold the Loom of Locke whose Woof rages dire,*
> *Washed by the Water-wheels of Newton: black the cloth*
> *In heavy wreathes folds over every Nation: cruel Works*
> *Of many Wheels I view, wheel without wheel, with cogs tyrannic*
> *Moving by compulsion each other, not as those in Eden, which*
> *Wheel within Wheel, in freedom revolve in harmony & Peace!*

Blake felt that the new science threatened to enslave the free human spirit. He warned us "to cast off Bacon, Locke, & Newton" (Blake 1969, 818), precisely Jefferson's "three greatest men" (see 2.1). Bluntly put, Blake's message is that "Art is the Tree of Life, Science is the Tree of Death" ("A Memorable Fancy").

A recurrent theme in literature is that science, in its reductive analysis, destroys the beauty and mystery of nature. An oft-cited expression of this belief is the following passage from "Lamia" by English poet John

Keats (1795–1821), a protest against the effect that Newton's analysis of the spectrum supposedly has on our appreciation of the beauty of the rainbow:

> *Do not all charms fly*
> *At the mere touch of cold philosophy?* [i.e., science]
> *There was an awful rainbow once in Heaven;*
> *We know her woof, her textures; she is given*
> *In the dull catalogue of common things.*
> *Philosophy will clip an Angel's wings,*
> *Conquer all mysteries by rule and line,*
> *Empty the haunted air, and gnomed mine—*
> *Unweave a rainbow . . .*

American poet e. e. cummings (1894–1962) offered a more contemporary expression of the distaste for science:

> *O sweet spontaneous*
> *earth how often have*
> *the*
> *doting*
> *fingers of*
> *prurient philosophers pinched*
> *and*
> *poked*
> *thee,*
> *has the naughty thumb*
> *of science prodded*
> *thy*
> *beauty*
> ("O sweet spontaneous," Cummings 1991, 58).

> *While you and i have lips and voice which*
> *are for kissing and to sing with*
> *who cares if some one-eyed son of a bitch*
> *invents an instrument to measure Spring with?*
> ("voices to voices, lip to lip," 262)

> *I'd rather learn from one bird how to sing*
> *than teach ten thousand stars how not to dance*
> ("you shall above all things be glad and young," 484)

What is the source of these passionate outcries against science? Is there any substantial justification for them? We don't have to look only to romantic humanists, reacting to science and technology with mistrust and distaste (and often ignorance), to find a bleak scientific view of nature. Many scientists and scientifically minded philosophers have expressed it themselves. The belief that science might shatter our ordinary world can be traced back to the very foundations of the new physics in the seventeenth century.

The ancient Aristotelian tradition divided motions into the categories of "natural" and "violent," or unnatural. In natural motion, the cause of the change is within the body; in violent motion, the cause is external. When a stone falls to earth, its motion is natural. The "final" cause of the stone's fall is its goal; the earth is the stone's natural place, to which it is drawn, and as it nears "home" it moves faster, more "jubilantly." In place of such teleological explanations, Galileo described the motions of bodies in terms of mathematical relations in space and time. In changing the question about motion from the "why" of purpose to the "how" of mathematical relationships, he banished purposes as causes or explanations of natural motion.

Since measurable mathematical quantities are the properties of objects by which their motion is now to be explained, the qualitative sensory qualities are not thought of as ultimately real in the new scientific description of the world. In a famous passage in "The Assayer" (1623), Galileo concluded his inquiry into the nature of heat thus: "People in general have a concept of this which is very remote from the truth. For they believe that heat is a real phenomenon, or property, or quality, which actually resides in the material by which we feel ourselves warmed" (Galileo 1957, 274). Galileo held that colors, tastes, smells, and sounds are likewise not really part of physical objects but exist only as sensations in the mind of the perceiver (or more exactly, as certain tiny shapes, sizes, and motions of particles that cause perceptions in human beings):

> To excite in us tastes, odors and sounds I believe that nothing is required in external bodies except shapes, numbers and slow or rapid movements. . . . I think that if ears, tongues, and noses were removed, shapes and numbers and motions would remain, but not odors or tastes or sounds. The latter, I believe are nothing more than names when separated from living beings, just as tickling and titillation are nothing but names in the absence of such things as noses and armpits. (276–277)

In this way, Galileo was led by his new physics to distinguish the qualities really present in physical objects from those that are merely subjective qualities of human sensation. This distinction between "primary" and "secondary" qualities has remained in use in science and philosophy right up to the present day.

To many people, the implication of science has seemed to be that "the really important world outside was a world hard, cold, colorless, silent, and dead; a world of quantity" (Burrt 1932, 239) or, to phrase it another way, "a dull affair, soundless, scentless, colourless; merely the hurrying of material, endlessly, meaninglessly" (Whitehead 1925, 55). A theory of atomism that was associated with Newtonian mechanics postulated that the ultimate parts of all things are atoms—minute particles that are indivisible and unchanging except for motion in space. The ancient Greek philosopher Democritus (c. 460–357 B.C.) had proposed an atomist view that reduced all phenomena to atoms moving in the void. Combined with atomism, the new physics became known in the seventeenth century as "the corpuscular philosophy": "The view became current that all the operations of nature, all the fabric of the created universe, could be reduced to the behaviour of minute particles of matter, and all the variety that presented itself to human experience could be resolved into the question of the size, the configuration, the motion, the position and the juxtaposition of these particles" (Butterfield 1957, 132).

In 1810, Johann Goethe (1749–1832), a great figure in German literature who tried to be both poet and scientist, reacted strongly against the science of the Enlightenment. In his *Theory of Colors,* Goethe described many years of research he spent trying to refute Newton's claim that white light is composed of the colors of the spectrum. Newton's analysis seemed to Goethe to sully the color white as a symbol of simplicity and purity. He claimed that he could *see* that perceived white is simple, not composite, and thus that our perceptual experience contains qualities to which no scientific analysis can do justice. (We sense white as an absence of color even though the sensation is caused by an aggregate of the frequencies of visible light.) Philosophers continue to debate the status of secondary qualities such as colors, but Goethe's efforts to establish a qualitative science had almost no impact on mainstream science in its quest to explain the physical world in terms of mathematically measurable quantities. Indeed, electromagnetic quantities—as the underlying nature of light—have been added to the original seventeenth-century list of primary qualities.

Science has thus seemed to divide our understanding of nature into two images: the scientifically described physical world and the world as presented in our sensory experience. Bertrand Russell, himself very much a champion of the scientific approach, once penned the following chilling passage on the position of humanity in the world that science seems to reveal:

That man is the product of causes which had no prevision of the end they were achieving; that his origin, his growth, his hopes and fears, his loves and his beliefs, are but the outcome of accidental collocations of atoms; that no fire, no heroism, no intensity of thought and feeling, can preserve individual life beyond the grave; that all the labours of the ages, all the devotion, all the inspiration, all the noonday brightness of human genius, are destined to extinction in the vast death of the solar system . . . all these things, if not quite beyond dispute, are yet so nearly certain that no philosophy which rejects them can hope to stand. . . . Such in outline but even more purposeless, more void of meaning is the world which Science presents for our belief. (1917, 56–57)

A number of scientists have also described the scientific image of the world in similarly bleak terms. French molecular biologist Jacques Monod wrote that

It is perfectly true that science outrages values. Not directly, since science is no judge of them and *must* ignore them; but it subverts [that] upon which the animist tradition . . . has made all ethics rest. . . . Now does man at last realize that, like a gypsy, he lives on the boundary of an alien world. A world that is deaf to his music, just as indifferent to his hopes as it is to his suffering or his crimes. (1971, 172–173)

Julian Huxley wrote that science presents "a universe of appalling vastness, appalling age, and appalling meaninglessness" (quoted in Otto 1940, 242). The physicist Steven Weinberg muses that "the more the Universe seems comprehensible, the more it also seems pointless" (1977, 144).

We cannot in this book explore all the deep philosophical issues raised by this image of science, but we shall say a little more about them in our concluding Chapter 10. For the moment, we should note that it is not the only view on offer. Among the alternatives are the images of science we consider next.

2.4 Utopias and Anti-Utopias: Science and Human Affairs

In the eighteenth century, various utopian thinkers, inspired by the success of the new science in astronomy and mechanics, proposed the application of scientific methods to study and control human behavior as well. A rather stark version of the philosophy of mechanism was presented by French Enlightenment thinker Baron d'Holbach (1723–1789) in his book *The System of Nature*. He was following up the worldview that Galileo and Descartes had foreshadowed in the previous century. Holbach embraced materialism and determinism as consequences of the new understanding of nature made possible by Newtonian mechanics. He viewed nature as a great machine and the human mind as nothing but a modification of the brain. (The most notorious expression of the materialist view is the pronouncement by Pierre Cabanis (1757–1808) that the brain secretes thought as the liver secretes bile (cited in Copleston 1960, 65). Freedom is an illlusion, d'Holbach argued: "If a man believes himself free, he is merely exhibiting a dangerous delusion and an intellectual weakness. It is the structure of the atoms that forms him and their motion propels him forward" (1868). With no freedom, there can be no sense to moral blame; crime is thus a kind of disease. Yet d'Holbach, curiously enough, offered a prescription for happiness: Learn to live in accord with reason!

Early nineteenth-century French thinkers Comte de Saint-Simon (1760–1825) and Auguste Comte (1798–1857) recommended a scientific approach to all questions in their "positive philosophy." Possessed of extraordinary ambition, they believed it possible to reconstruct the whole of human society on a rational, scientific basis. Saint-Simon proposed what now seems a pretty ridiculous scheme for the founding of the Religion of Newton, with universal gravity as "the sole cause of all physical and moral phenomena." He seems to have believed that society, politics, and human life in general were all reducible to gravitational attractions. By means of the new religion of science, operating through a "Supreme Council of Newton" to educate the public, the human race might perfect society and arrive at a new Golden Age. Here was an explicitly religious faith in science, with a deified Newton playing a Christlike role as Science incarnate in human form!

Many people have felt that such rationalistic idealism of enthusiasts for scientific progress ignores the dark side of human nature. Russian novelist Fyodor Dostoevsky (1821–1881) gave tortured expression to this fear in his story *Notes from Underground,* in which the miserable protagonist takes as a symbol of the brave new society the Crystal Palace, the huge structure built in 1851 to house an exposition in London of new industrial products. The palace was itself a triumph of new technology in architecture, but the Underground Man finds it and its contents a dreadful bore. His chief concern is the potential of scientific theories to predict behavior, which he fears will transform each of us "from a human being into an organ stop" (1960, 24).

One might worry about another supposed consequence of scientific progress: If scientific knowledge increases our freedom, it also heightens our burden of responsibility. A theme of existentialist philosophy is the anxiety caused by such burdensome freedom—the greater the choice, the greater the anxiety. The flood of new information in science and new techniques and devices in technology may result in "future shock," a difficulty in adapting to the many new choices that science and technology offer us. In the words of the book of Ecclesiastes, "In much wisdom is much vexation, and he who increases knowledge increases sorrow" (1:18 RSV). The more we know, the fewer things we can call "acts of God" (natural disasters that are beyond human control).

The scientific mentality of detachment and objectivity and "technological rationality" with its values of efficiency and control have been blamed for many evils. Litanies of the baleful influences of science and technology on our culture are often recited. Some people believe that science has divorced us from nature and that it has regimented and routinized labor, making workers adjust the natural rhythms of the human body to the relentless pace of machines. For example, Friedrich Juenger complained of "the disintegration of nature by scientific thought" (1949, 107) and the subjection of men by scientific technology "to a mechanical, lifeless time" (48). Herbert Marcuse has written about how reality becomes flat and dreary under the repression of technological "rationality" (1964).

These are the kind of fears embodied in works describing technological anti-utopias (or *dystopias*) such as Yevgeny Zamyatin's *We*, Aldous Huxley's *Brave New World*, B. F. Skinner's *Walden Two,* and George Orwell's *1984*. These dystopias have in common a central political control that represses individual freedom. Life is regimented, mechanized, and "unnatural," with behavior controlled either by force or more subtly, as in Huxley's world, by psychological conditioning. The ultimate goal in such societies appears to be stability and maintenance of power by those in control. But there is an alternative, more positive, view of the social role of science, which we must consider next.

2.5 The Two Cultures: Science as Humanizing?

Sir Charles Percy Snow stirred up an interesting controversy about the relation of scientific and humanistic understanding in his Rede Lecture of 1959, "The Two Cultures." He claimed that "the intellectual life of the whole of Western society is increasingly being split into two polar groups. . . . At one pole we have the literary intellectuals. . . . at the other scientists. . . . Between the two [there is] a gulf of mutual incomprehension—sometimes (particularly among the young) hostility and dislike, but most of all lack of understanding" (1959, 4). Snow thought he could straddle the gap between these cultures that he distinguished so dramatically; he described himself as "by training I was a scientist; by vocation I was a writer [of novels]" (1959, 1). Few have been happy with his assessment of two divergent cultures within our society. But the amount of controversy he aroused shows that he struck a nerve, and his phrase "the two cultures" has been widely used.

Snow's distinction reflected in part the tradition of British education, which until recently endorsed a curriculum dominated by the humanities, especially the Latin and Greek classics. The only sciences included in the traditional liberal arts in medieval times were mathematics and astronomy. The rest had to struggle to gain recognition in the universities in the nineteenth century (see 1.3). The interesting question raised by this split in priorities is not so much about different cultures but about differences in individual attitudes toward science, which seem to fall into a fairly coherent dichotomy. Historian of science George Sarton had noted that "the most ominous conflict of our time is the difference of opinion, of outlook, between men of letters, historians, philosophers, the so-called humanists, on the one side, and scientists on the other" (1936, 54). Thus, Snow's two cultures are better viewed not as classes of individuals (scientists versus humanists) but as contrasting attitudes that the same person might feel in different contexts. As Jacob Bronowski (1965) rightly protested, we can't categorize some people as pure thinkers and others as pure "feelers."

The following is a list of opposing qualities that characterize the contrast between the stereotypical scientist and the traditional literary, artistic, romantic person:

reason	feeling
abstract	concrete
general	particular
restrained	spontaneous
determined	free
logical	intuitive
simple-minded	muddle-headed
analytic	synthetic
atomistic	holistic
reality	appearance
optimistic	pessimistic
masculine	feminine
Yang	Yin
left brain	right brain

Such black-and-white contrasts shouldn't be taken too seriously—they tend to break down in particular cases—but the stereotypes do stand out rather clearly. (A similar set of polar opposites is presented in Robert Pirsig's *Zen and the Art of Motorcycle Maintenance* [1974], which tries to reconcile the classic and romantic viewpoints.) This polarity is a pervasive theme in discussions of the role of science in society and is expressed in many guises. For example, consider the complaint against the scientific culture as having "too much left brain/technological/ number-crunching/logical stuff, and not enough of the fey, intuitive, creative, feeling, lucky, right-brain bit. That is the root of our current problem" (Kinsman 1993, 56). Consider also the following contrast between these two worldviews:

> Western culture has for a long time been trying to manage a society based on two incompatible pictures of reality—one 'scientific' and the other in some sense 'spiritual'. . . . The scientific worldview was, and is, powerfully influential in the political, economic, and industrial institutions of modern society. However, people have tended to live their lives by, and society to derive its values from, the other, competing worldview. (Harman 1993, 140)

Snow complained that the typical literary intellectual did not know much about science and, worse yet, was not aware of the depth of his ignorance. An example to support Snow's point is this passage by Anglo-French writer and poet Hillaire Belloc, who offered an unintentionally naive caricature of how science works:

> Anyone of common mental and physical health can practise scientific research. . . . Anyone can try by patient experiment what happens if this or that substance be mixed in this or that proportion with some other under this or that condition. Anyone can vary the experiment in any number of ways. He that hits in this fashion on something novel and of use will have fame. . . . The fame will be the product of luck and industry. It will not be the product of special talent. (1931, 226)

Presumably, Belloc believed that mere perseverance and good fortune are required for success in science, as opposed to the special talent needed to produce literary works. True enough, scientific discoveries generally involve patient experimenting and in some cases a bit of luck. But there is more to science than that—one must have imagination and inspiration. And although prolific U.S. inventor Thomas Edison (1847–1931) assigned only 1 percent of the credit for scientific discoveries to inspiration, as compared to the 99 percent he attributed to perspiration, a glance at actual cases in the history of science demonstrates just how crucial that 1 percent is.

Snow, in sharp contrast to the proponents of pessimistic images of science whom we have considered in 2.2 through 2.4, was distinctly more positive about scientific culture and attitudes. His general view was that the future belongs to science (whatever future humankind has, that is), and he argued that science-based industrialization "is the only hope of the poor" (1959, 27) since "health, food, education; nothing but the industrial revolution could have spread them right down to the very poor. These are primary gains—there are losses too, of course. . . . But the gains remain. They are the base of our social hope" (1959, 29).

Other thinkers also find hope for the future in the scientific attitude. Some see the scientific community as providing a good model for democratic freedom and cooperation. Robert Cohen suggested that the ethic of the scientific community is "the democratic ethic of a cooperative republic" (1974, 141) and that the tradition of science has achieved a "unique fusion of obligation and rebellion" (140), of competition and yet cooperation. Science tends to foster a democratic forum of ideas, with an obligation to make the results of scientific research public and not to repress contrary views. Of course, such an ideal has not always been met, as will be demonstrated in some of our case studies. But science has flourished best where its traditional values are recognized— and science perhaps contributes to their recognition: freedom, tolerance, independence, perseverance, originality, truthfulness, and international fraternity. Has any other institution actually exhibited a better model of human behavior? J. Robert Oppenheimer argued optimistically that "of all intellectual activity . . . science alone has turned out to have the kind of universality among men which the times require" and that "if one looks at past history, one may derive some encouragement for the hope that science, as one of the forms of reason, will nourish all of its forms" (1986, 208–209).

Recently, the distinguished biochemist Max Perutz has echoed not only the Baconian spirit detailed in 2.1 but also Snow's idea of the positively humanizing mission of science: "Christians throughout the Dark and Middle Ages seem to have interpreted this injunction [Jesus' Sermon on the Mount] to mean that man should not strive to better his lot in this world but should prepare himself for the next. Science has reversed these values by convincing man that it lies in his power to improve the conditions of his own life and that of his fellow men in this world" (1989, 216). Perutz claimed not just that science has given us knowledge about the natural world, which is applicable for our own benefit, but that it has had a fundamental influence in changing human attitudes for the better. Science is here credited with having a civilizing or humanizing effect, a replacement of unrealistic religious values by robustly practical, this-worldly concerns. This is missionary zeal for science, indeed!

2.6 The Thesis of the Value-Neutrality of Science

Can science give us objective answers to questions about values? A common response to the whole question of how science relates to values is to say that it is fundamentally and strictly value-neutral. Science, so this argument goes, neither invalidates all human values (as suggested in 2.3) nor supports particular values, whether "good" or "bad" (as suggested in 2.2 and 2.5). According to this view, science can deal only with facts, rather than values; with techniques, rather than goals; and with means to ends, rather than ends in themselves. The question of how to use the power that scientific understanding gives us, to what ends to apply the new techniques for changing the world that science offers us, is completely up to us. To invoke the commonly used phrase, all such questions are "for society to decide." The following three claims can be distinguished as part of this image of scientific neutrality:

1. Science offers us objective knowledge of how the world works and hence of the consequences of various interventions in it, but there can be no such knowledge of whether we *should* make any particular interventions. A sharp distinction between facts and values is here assumed to rule out any knowledge of the latter, so the adoption of goals and policies is seen as a matter of merely "subjective" individual opinion. This separation of facts from values seems to follow from the impossibility, as Scottish philosopher David Hume (1711–1776) said, of deriving "ought" from "is"—a conclusion expressing a value judgment can only be deduced from valuation assumptions. Thus, well-known philosopher of science Carl Hempel argued that although hypothetical judgments like "If you administer 5 mg of cyanide to this baby, it will die" express scientific facts, categorical judgments of value such as "Killing babies is evil" are "not amenable to scientific test and confirmation or disconfirmation" (1960, 45). According to this view, science provides means ("instrumental" values) but never ultimate ends. As German economist/sociologist Max Weber (1864–1920) remarked, science is like a map that can tell us how to get to many places but not where to go.

2. The only value recognized by the scientist as such, the only value actually involved in doing science, is that of knowledge for its own sake. She may welcome the possibility of useful applications of her research, but as a scientist she is devoted purely and simply to the extension of human knowledge as an end in itself. In this respect, she is just like her university colleagues the Sanskrit philologist, the medieval historian, and the pure mathematician, who delight in the achievement of new knowledge and understanding even if there is no prospect of its being useful in any practical way. As an eloquent nineteenth-century expression of this view, consider this passage from the physicist Hermann von Helmholtz:

Whoever, in the pursuit of science, seeks after immediate practical utility, may generally rest assured that he will seek in vain. All that science can achieve is a perfect knowledge and a perfect understanding of the action of natural and moral forces. Each individual student must be content to find his reward in rejoicing over new discoveries, as over new victories of mind over reluctant matter, or in enjoying the aesthetic beauty of a well-ordered field of knowledge, where the connection and filiation of every detail is clear to the mind, he must rest satisfied with the consciousness that he too has contributed something to the increasing fund of knowledge on which the dominion of man over all the forces hostile to intelligence reposes. (Helmholtz 1893, quoted in Ravetz 1971, 39)

3. The applications of scientific knowledge are for "society" to decide. The applied scientist or technologist is the servant of other people, using his expertise toward ends that are chosen in whatever way it is that individuals and institutions make such choices. Some believe that there is such a thing as knowledge of the right way for human beings to live—whether derived from an immemorial social tradition, a sacred book, an authoritative church, a theocracy of mullahs, the ideology of a ruling party, or an inspired leader. All these different believers can equally apply scientific knowledge toward their various (and perhaps conflicting) visions of what constitutes the good life for humanity. Those who are more skeptical about the possibility of *knowledge* of values will typically suggest at this point that the ends to which scientific knowledge is to be applied should be determined by a democratic process that allows society's decisions to emerge from the mass of individual, subjective opinions. U.S. physicist Edward Teller, whom we will meet in 9.1, has expressed this view as follows:

The scientist's responsibility is to find out what he can about nature. It is his responsibility to use new knowledge to extend man's power over nature. . . . When the scientist has learned what he can learn and when he has built what he is able to build his work is not yet done. He must also explain in clear, simple, and understandable terms what he has found and what he has constructed. And there his responsibility ends. The decision on how to use the results of science is not his. The right and duty to make decisions belongs to the people. . . . The scientist has done his full duty only if he becomes a full participant of our vital, paradoxical, multicolored, democratic society. (1960, 21–22)

There are splendid-sounding words in all these quotations, and they express much of the conventional wisdom of our age. Yet all three elements of this picture of the value-neutrality of science are open to question. Thesis 1 makes a controversial assumption about a sharp distinction between facts and values and in so doing raises profound philosophical questions that we cannot hope to explore in any depth in an introductory book like this. However, in Chapter 10 we shall examine a little more systematically the question of how science and values are related. And in Chapters 4 through 9, as we survey a number of case histories in science, bringing out the variety of human motivations and the social pressures involved in them, we will find plenty of reasons to question theses 2 and 3.

Suggestions for Further Reading

Ben-David, H. 1971. *The Scientist's Role in Society.*
Berman, M. 1981. *The Reenchantment of the World.*
Burrt, E. A. 1932. *The Metaphysical Foundations of Modern Science.*
Cornelius, D. K., and E. St. Vincent. 1964. *Cultures in Conflict: Perspectives on the Snow-Leavis Controversy.*
Farrington, B. 1966. *The Philosophy of Francis Bacon.*
Gendron, B. 1977. *Technology and the Human Condition.*
Merchant, C., 1980. *The Death of Nature: Women, Ecology, and the Scientific Revolution.*
Passmore, J. 1978. Science and Its Critics.
Pirsig, R. 1974. *Zen and the Art of Motorcycle Maintenance: An Inquiry into Values.*
Rifkin, J. 1991. *Biosphere Politics: A New Consciousness for a New Century.*
Shelley, M. W. 1994. *Frankenstein, or the Modern Prometheus.*

What Motivates Scientists?

To gain a more detailed perspective on science and its historical development, we shall be examining in this chapter the work of some selected scientists in their various historical, social, political, and economic contexts. We shall ask how pure scientific curiosity, the valuation of knowledge as worth having for its own sake, interacts in the real world with other motives, values, and pressures. We shall look at some famous (and notorious) examples as well as some lesser-known figures; unfamiliar aspects of certain great scientists will also be considered. We will be asking questions about what motivated them to do what they did in the circumstances in which they lived and worked. In this way, we shall try to bring out the values that have actually been involved in scientific work and to suggest how they have developed and changed over time. Our selection and presentation of case studies is bound to reflect our own presuppositions of what science is and ought to be. It is our hope, however, that the large range of examples, themes, and issues we raise will avoid a narrow bias.

3.1 Who Counts as a Scientist?

If we are to focus on particular scientists, we need first to address the question, Who counts as a scientist? Everyone has heard of great scientists such as Galileo, Newton, Darwin, and Einstein. But besides such towering figures, there are lots of others who obviously count as scientists. Many thousands of people presently consider themselves part of the scientific community in some way. Some of them, though hardly belonging to the same class as Einstein, have made important contributions to scientific theory, but many of them have not—if by *important* we mean "generally acknowledged as significant by scientists in the relevant discipline."

It has been claimed that more than 90 percent of all the scientists who have ever lived are alive today. Whatever the exact definition of *scientist,* this assertion points up something special about our age: Many more people are employed in some aspect of science than ever before. If we were to include all those whose occupations make use of scientific research in some way—computer technologists, engineers, physicians, and laboratory workers, for example—then the total could be as high as several million scientists in the United States alone. But even though engineers and physicians have to study science as part of their training, use science in their work, and may make contributions to science, they do not commonly call themselves scientists. Likewise, the National Science Foundation in the United States classifies only a small percentage of physicians and engineers as scientists. Laboratory technicians, who may or may not be required to understand the theories behind the research to which they are contributing, are not so much players in the game of science as they are assistants on the sidelines. Science teachers, who may have much scientific knowledge but are not themselves trying to add to it, are also not usually considered scientists.

Thus, we must remember the distinction between the millions of people whose work is connected with science in some way and those who are scientists, doing research that attempts to extend scientific knowledge and understanding. Having and using a store of scientifically obtained knowledge, however accurate and useful it may be (such as the databases of the U.S. Departments of Defense, Energy, or Agriculture), is not sufficient to make one a scientist. As we have seen in Chapter 1, the very concept of the scientist has evolved over time and may still be changing. Despite the blurring of the concept today, with the industrialization of science and the closer association between science and technology, we can usually single out clear cases of research work that have contributed directly to new knowledge. So we can distinguish basic (or "fundamental" or "research") scientists as those attempting to do creative work in science, trying in some way to enlarge knowledge or improve on existing scientific theory. But it has been difficult for those who were not gentlemen (in the early years of modern science) or not male and white (at least until recently) to become accepted as scientists. The practice of science—and the social status of scientists—has never been immune to wider social influences and pressures.

Besides those who use science without attempting to contribute to it, there are also those who may *claim* to be scientists but who do not deserve this title. The primary criteria for identifying pseudoscience, signs by which the bogus can be distinguished from the genuine article, relate to the logic and methodology of science. The pseudosciences generally fail to give any novel, and definitely falsifiable, observational predictions (see 1.1). Other secondary criteria have to do with the social relations of would-be scientists and the motivations for their work. Since scientific communities are in large part defined by their institutions—

their meetings, publications, and networks of communication—isolation from the scientific community tends to mark certain theorists as pseudoscientists. But such criteria would be circular if we had no independent way to pick out genuinely scientific communities. (Why don't astrologers form such a community?) Once again, we have to ask whether members of a would-be scientific community systematically submit their theories to the test of observation.

A kind of severe honesty is built into proper scientific inquiry. Individual scientists do not always live up to the ideal, but at least in mainline academic science, the honesty of scientists compares rather well with other professions. They may not be innately better human beings, but the nature of scientific inquiry, scientific communication, and the acquisition of scientific reputation require that all experimental claims be in principle replicable by anyone anywhere (provided one has the requisite material and equipment). Fraudulent claims are unlikely to advance an individual scientist's career for very long. But let us now ask what motivates people to engage in the stringent mental discipline that is involved in doing science.

3.2 The Variety of Motives Driving Scientists

John Ziman noted in his oft-cited work *Public Knowledge* (1968) that our image of the scientific life is "strongly colored by the mythology of the [heroic] period," with the scientist pictured as "a lonely, dedicated personality, grappling with problems that he has set himself," who is inner directed and follows his own star (82). As the case studies in Chapter 4 will illustrate, this image of the scientist as tireless seeker after truth is not entirely off the mark for many of the great classic scientists, though the way science works is not always so simple.

Suppose little Sonia says, "When I grow up, I want to be a scientist!" To what image of scientists is she likely to be attracted? What draws a young man or woman into science? Many scientists have remarked that they were inspired by reading about a great discovery that struck a deep chord of interest. A number of other possible motivations come to mind: the desire to be recognized, to acquire prestige or fame; the desire to help people, to improve the human condition in some way, and to be recognized as useful. But what stands out in many scientists' own accounts of their motivations is curiosity to see how nature works and to discover something that no one has known before.

People's motives tend to appear more diverse and complex the more carefully we scrutinize them. If Mr. Jones tells us why he became a lawyer, a businessman, or a psychotherapist, we might wonder what his real reason was, what his underlying motives were. A Freudian psychologist might suggest one kind of motive; a behaviorist would offer a different account. What brings someone into science in the first place, and what actually motivates a scientist at any given moment as he or she works in the lab or in the field? At one level we can identify the general aims of the enterprise: to advance theoretical knowledge or to make discoveries useful to humankind. But the immediate, short-term motivations that scientists report may be curiosity about a particular phenomenon, enjoyment of tinkering about in the lab, the need to finish a paper to impress the boss or an appointments committee, or the pressure to meet the deadline for a grant application.

Some of the possible motivations that come to mind for choosing any occupation are that the work is fun or glamorous, that the work helps people or satisfies parental expectations, and that it promises fame, power, or money. J. G. Crowther, in his pioneering book *The Social Relations of Science* (1941), offered a list of five "personal motives" for scientists doing research that include most of those usually cited:

> The one which is best known, and most frequently announced by scientists themselves, is *curiosity* or the desire for understanding for its own sake. Another very powerful and general motive is the desire for *reputation*. A third is the need to *earn a living*. A fourth is the desire to *enjoy oneself*. A fifth is the desire to *serve humanity*. Very little psychological research has been done to discover the relative weight of these motives in practice. (511, our italics)

Without attempting to offer anything that could be dignified with the title of psychological research, we shall in the chapters that follow present some case studies that illustrate the operation of these motives as well as others. We shall try to assess how personal motives and social influences have driven science in the past and, with some differences, how they continue to drive research today.

Looking back again to Bacon's vision of science as discussed in 2.1, we find a very definite view of what the motivation of scientists *ought* to be. Knowledge—that is, science—should not, in Bacon's opinion, be pursued "either for pleasure of the mind, for contention or for superiority to others, or for profit, or fame, or power, or any of these inferior things; but for the benefit and use of life" (1870, vol. 4, 21). In this severe judgment, Bacon showed a shrewd awareness of the mixed nature of human motives. But what may strike the contemporary reader as strange is the implication that "pleasure of the mind," which we might describe in more dignified terms as pure intellectual satisfaction in understanding the workings of nature, is an *inferior* motive. This clashes with the primacy most scientists today give to the desire for understanding for its own sake.

Bacon's extreme emphasis on utility is, however, more understandable if taken in its historical context. Writing at the threshold of the rise of modern science, Bacon felt it necessary to oppose the prevailing tradition that the pursuit of knowledge of nature was proper only for gentlemen who had the ability and the leisure for theoretical speculation. The practicalities of life, such as cookery and childrearing, and the physical labor involved in mining, building, and farming were largely the work of slaves in ancient Greece and of wives, servants, and workers in Bacon's time. But he wished to persuade his readers that scientific knowledge of nature would pay practical dividends in raising the standard of living of everyone in society, not just the elite. And he wanted knowledge to be publicly available rather than reserved for the initiates of secret societies of astrologers, alchemists, or magicians; claims to knowledge should be open to test by any qualified observer. So, in tune with the rising democratization of economic opportunity, Bacon focused on "the benefit and use of life" as the proper aim of science.

Among scientists themselves, more emphasis has usually been given to the Greek ideal of knowledge for its own sake as the purest and loftiest motivation for scientific work. Aristotle proclaimed that "all men by nature desire to know" (1984, 980a22). Since Bacon's time it has become a standard view among scientists that pure curiosity should be the primary—and perhaps the only—motivation of the true scientist. Most scientists (at least until recently) would probably agree with Helmholtz's expression of this view (see 2.6) and with the similar view of French mathematician and physicist Henri Poincare (1854–1912), who claimed that "the scientist does not study nature because it is useful to do so. He studies it because he takes pleasure in it, and he takes pleasure in it because it is beautiful" (1952, 22). Another example of this orientation is provided by Glenn Seaborg, Nobel Prize winner in chemistry: "In his basic research, the motivating force is not a utilitarian goal; its keynote is intellectual curiosity and a desire for discovery" (Love and Childers 1965, 33).

Much of the knowledge gained through scientific research has nonetheless yielded useful applications. In our own time, it is a common expectation that advances in theoretical science will yield useful technology as a by-product. Yet however strongly they are hoped for, the spin-offs from fundamental scientific research cannot be confidently predicted in advance. But since those holding the governmental and industrial purse strings for expensive research projects (like the Hubble telescope or the supercollider) are usually motivated by considerations of utility or profit, many scientists today need to play up the probable applications of their research in order to get funding for it. Most scientists as individuals, however, still tend to emphasize the aim of adding in some way to our understanding of nature.

Einstein was one of the purest examples in all science of sheer intellectual intoxication with understanding the deepest secrets of the physical universe. His expression of this primary motivation for the practice of science often takes on a religious tone, as in the following remarkable passage:

> Many kinds of men devote themselves to Science, and not all for the sake of Science herself. There are some who come into her temple because it offers them the opportunity to display their particular talents. To this class of men science is a kind of sport in the practice of which they exult, just as an athlete exults in the exercise of his muscular prowess. There is another class of men who come into the temple to make an offering of their brain pulp in the hope of securing a profitable return. These men are scientists only by the chance of some circumstance which offered itself when making a choice of career. If the attending circumstance had been different they might have become politicians or captains of business. Should an angel of God descend and drive from the Temple of Science all those who belong to the categories I have mentioned, I fear the temple would be nearly emptied. But a few worshippers would still remain. . . . Let us fix our gaze on those who have found favour with the angel. For the most part, they are strange, taciturn and lonely fellows. . . . What has led them to devote their lives to the pursuit of science? Personally I am inclined to agree with Schopenhauer in thinking that one of the strongest motives that lead people to give their lives to arts and science is the urge to flee from everyday life, with its drab and deadly dullness, and thus to unshackle the chains of one's own transient desires. . . . But to this negative motive a positive one must be added. Human nature has always tried to form for itself a simple and synoptic image of the surrounding world. (from the preface to Planck 1936)

Einstein went on to describe in more detail the particular kind of worldview (*Weltanschauung*) that theoretical physics forms and why it is worth seeking. While making very clear that he regards the desire for this kind of understanding as the most fundamental and admirable motive for doing scientific research, he admitted that "many worthy people" who have been motivated by fame or profit have built "a great portion of the Temple of Science." It is striking that he did not even mention Bacon's supreme motive for scientific work—the hope for practical usefulness. This might be thought to fall under the rubric of "a profitable return," but usefulness is distinguishable from financial reward, since it is quite possible to benefit others without making money thereby (for example, nonprofit charities) or to make money without benefiting others (as in drug dealing). In Chapter 6, we shall cite examples of scientists motivated at least partially by the desire to be useful to industry, to their nation, or to humanity in general.

One element that will show up in many of our case studies is the intensity of the motivation required to make any substantial achievement in science. Why have scientists often been willing to persist in frustrating and exhausting toil—to achieve what? There is an apparent paradox in scientists passionately pursuing what is supposed to be an objective, dispassionate quest for knowledge. But scientific research clearly requires emotional drive as well as intellectual ability and discipline. When we examine some of the historical details, it is striking how difficult it was to achieve scientific discoveries that in retrospect may seem relatively obvious. Everybody knows that the heart circulates the blood, so why was William Harvey's inquiry so long and tortuous? (See 6.1.) Why was it so difficult to realize that combustion is oxidation? (See 4.3.) Or to see heat as a kind of energy? Why was it only in the late nineteenth century that the bacterial cause of many diseases was widely recognized, especially since microscopes had been available for over two hundred years? (See 6.1.) What motivated certain people to struggle so hard to solve such problems?

What do scientists actually do? Michael Faraday summed up a less than glamorous image: "Work, finish, publish" (quoted in Mackay 1991, 56). Oliver LaFarge, in an article entitled "Scientists are Lonely Men," offered a description that fits much of the day-to-day work of scientists: "To an outsider who does not know of (the) emotion, the scientist suggests an ant, putting forth great efforts to lug one insignificant and apparently unimportant grain of sand to be added to a pile, and much of the time his struggle seems as pointless as an ant's" (Shapley et al. 1965, 35).

On his own work, LaFarge commented, "I spent hours, deadly, difficult hours, extracting lists" (to solve an abstruse problem in linguistic archeology) (36). Successful scientists tend to be workaholics; their work is the most important thing in their lives. J. Robert Oppenheimer could think of only one quality he could ascribe to "all good physicists. . . . They all give a lot for physics; they care about it; they live it; they breathe it; they respect it" (Love and Childers 1965, 44).

A clue to the basic motivation of scientists is found in descriptions of their reactions to a discovery. The classic example is the famous story of ancient Greek mathematician and inventor Archimedes (see 4.1) leaping from his bath and running naked through the streets exclaiming *"Eureka!"* ("I have found it!"). He had hit on what is now known as Archimedes' principle: A body submerged in a fluid loses weight equal to that of the volume of fluid displaced. Subsequent scientists have at times used florid terms to describe the thrill of the scientific quest and the delights of discovery. When the attention of Italian physiologist Luigi Galvani (1737–1798) was drawn to the phenomenon of a frog's leg twitching from electric stimulation (leading to Alessandro Volta's [1745–1827] invention of the electric battery), he wrote, "I was inflamed with an incredible zeal and eagerness to test the same and to bring to light what was concealed in it" (Magie 1963, 422). James Bonner, a botanist, noted the "moments of greatest exhilaration, the moments when you suddenly discover that you have found out something new, something that no one has ever known before," which make the drudgery worthwhile (Love and Childers 1965, 157). George Beadle (Nobel Prize winner in physiology and medicine) remarked of his own first discovery in science, "I rode the clouds for weeks"—and this because of "a small finding—the identification of a gene in Indian corn that controlled the pairing of chromosomes" (5).

Curiosity is the primary motive for being involved in science cited by scientists themselves, as Crowther noted. Scientific curiosity is of a very special kind; it seeks to understand how nature works. It is "objective" or impersonal in the sense that it does not presuppose that nature works in accord with our personal needs. Scientists ultimately assume that nature "makes sense," that there is some kind of order to be discovered in nature's workings. A kind of childlike wonder commonly drives scientists, as Newton's oft-quoted remark illustrates: "I do not know what I may appear to the world; but to myself I seem to have been only like a boy playing on the seashore, and diverting myself in now and then finding a smoother pebble or a prettier shell than ordinary, whilst the great ocean of truth lay all undiscovered before me" (Brewster 1855, vol. 2, ch. 27). Many great scientists seem naturally to have followed the advice of English biologist T. H. Huxley (1825–1895) to "sit down before fact as a little child" (1900, 219).

Scientific research can attract people who enjoy exploring the unknown. The adventure lies in the uncertainty; a scientist, not unlike a prospector for gold, hopes to find something wonderful. Behavioral psychologist B. F. Skinner seriously proposed that scientific activity can be explained similarly to gambling: The schedule of reinforcement tends to be relatively sporadic, but that very randomness—and the high degree of reinforcement when it does happen—makes scientists persevere in their quest. Scientific curiosity is never wholly satisfied. There are always new things waiting to be discovered, and there is always the hope of hitting another jackpot.

We can roughly distinguish three categories of motivations among scientists: purely individual motives, motivation directed toward bettering the scientific community, and motivation directed toward bettering society in general. The most basic motive—the one that gets people into science in the first place—is, as we have seen, *intellectual curiosity*. Closely associated with this is pleasure derived from the practical activity of doing science, which often amounts to *enjoyable tinkering;* in this sense, a scientist is not unlike a child taking apart a clock to see what makes it tick. The second kind of motivation is crucial for the development of science, namely, desire for *reputation and influence within the scientific community.* This can be distinguished from motivation of the third kind, which is fueled by an ambition to provide wider *public usefulness or benefit* or attain *fame, influence, or power* and which may not relate very closely to strictly scientific reputation.

The reputation that primarily concerns scientists is esteem among the scientific community—or at least in their particular discipline. Scientists like to share their results with those who will appreciate them. They typically hope to have their claimed discoveries recognized as such, to be given credit for them, and thus to acquire a reputation for good scientific work. Now that science has been professionalized, to be recognized as a scientist requires having one's work published in recognized scientific journals. One measure of success in science is how often other scientists cite one's publications. Given this social element—which is essential to the practice of science as we know it—it can become difficult to distinguish clearly between pure curiosity and the self-centered desire for reputation. We will examine some cases to illustrate this problem in Chapters 4 and 5.

Traditionally, most scientists have not sought public fame or power, in contrast to their avid and sometimes rancorous struggle for reputation and even dominance among their scientific fellows. (Scientists may claim to be humble before nature, but they are not always so humble before their colleagues.) In our time, however, there has been an increasing tendency for scientists to get involved in politics and public policy, often out of a sense of responsibility caused by the ever-increasing impact of science and technology on society. We will examine some cases in Chapters 6, 8, and 9. Today many scientists are mainly involved in applied work, which is directed toward useful results (see Chapter 6). Some scientists have enjoyed public fame (see 5.3), and some exercise influence on public policy, whether behind the scenes in lobbying and advising or in open campaigning (see Chapter 9).

The need to seek funding has increased greatly during the twentieth century, given the increased costs of so much scientific research. Even the need to "make a living" is rather different now; the expected standard of living is quite high in developed countries. And the temptations to make large profits from the commercial exploitation of scientific innovations is vastly greater today, especially in the new field of bioengineering. We will look at some examples of the relationship between science and money in Chapter 7. Prominent scientists can sometimes increase both their wealth and their fame by writing popular books about their exploits (see 5.1).

The following is our refined list of motives and influences that drive scientists' behavior, divided into three categories:

1. Motives internal to the process of scientific research:

~Scientific curiosity
~Pleasure in the process of doing research

2. Motives directed toward the scientific community:

~Desire for scientific reputation
~Desire for influence within the scientific profession

3. External influences on scientific research:

~The attractions of public fame
~Desire to find useful applications of scientific knowledge
~The need for funding (getting money for doing science)
~Desire to make profits from applying scientific research (making money from science)
~Ambition to influence public policy (behind the scenes or by public campaigning)

This rough categorization of motives and influences will be used to shape our presentation of the examples that follow in Chapters 4 through 9. We hope to illuminate the momentous issues we have raised about science and society not by means of an academic essay in philosophy, sociology, politics, or history but by means of a series of case studies chosen to illustrate important themes. But we must emphasize that the actual motivation for an individual scientist may be quite complex, usually involving several motives at once and varying according to the situation. Most scientists have to make a living from their work, and like most people they feel the attractions of rewards, whether of money or of reputation and status; sometimes they may hope to contribute to the welfare of humankind. Many scientists also desire positions of influence or power as the heads of university departments or governmental agencies, directors of laboratories, chairs of research councils, or government advisers.

Nevertheless, there is a clear sense in which scientific curiosity, the desire to understand nature, is primary. As Einstein noted, the other motives that we have cited can be found in many other kinds of human activity such as business, sports, and art. And the motives of usefulness, fame, power, and profit take special forms within science. Intellectual curiosity may not be sufficient to make one a scientist. Not every eccentric theorist about nature counts as a scientist; earning that title requires the discipline of submission to criticism from the scientific community. In different parts of the quest for discovering and applying knowledge, different motivations come to the fore. But no activity can count as scientific unless it answers to the distinctive desire to know and understand the truth about some aspect of nature.

Suggestions for Further Reading

Barnes, B. 1985. *About Science.*

LaFollette, M. C. 1990. *Making Science Our Own: Public Images of Science 1910–1955.*

LaTour, B., and S. Woolgar. 1979. *Laboratory Life: The Social Construction of Scientific Facts.*

Root-Bernstein, R. 1988. *Discovering.*

Ziman, J. 1968. *Public Knowledge: The Social Dimension of Science.*

4 Intellectual Curiosity and Experiment

We have cited intellectual curiosity as a primary motive for doing science, but how does that differ from other forms of curiosity? Human beings are curious in various ways. One kind is the frivolous, often prurient inquisitiveness indulged in by gossip columns, television talk shows, and the tabloid press. There is also simple curiosity to know any sort of fact, say, which team won the World Series in 1923—the kind of knowledge rewarded in quiz shows. But our concern is the distinctively *scientific* sort of curiosity, the desire to understand nature. This need not always be taken to the depth of quasi-religious passion that Einstein extolled and exemplified (see 3.2). His was a particularly dramatic and fundamental form of a longing that most of us have experienced, at least to some degree.

Scientific curiosity takes two main forms, which can be combined in various proportions: (1) the quest for laws or general patterns of events and (2) an interest in mechanisms, in the ways that things are constructed and operate. These are typically expressed in the desire to reveal mathematical structure or to discover constituents and causes. Aptitude for science can be identified early in life not only in young people who are intellectually excited by the certainty of mathematical proofs, for example, in geometry, but also in those (not always the same individuals) who enjoy taking things apart to find out what makes them tick. The abilities to understand and explain and to intervene and control, though often connected, do not always come together. Some scientists are supreme theoreticians but are notoriously inept at any practical craft, including the experimental testing of their own theories; others have a strong practical bent but only a modest grasp of theory. Most scientists combine these talents to varying degrees.

We shall look first at some scientists who were primarily theoreticians, using mathematics as a key to achieving fundamental understanding of nature; second, at some who supported their theoretical contributions by careful experimentation; and third, at some who were better at experimenting than theorizing, who primarily enjoyed the exercise of their practical skill.

4.1 Seeking Mathematical Patterns in Nature

Distinctively scientific curiosity about the structure of nature cannot be fully appreciated apart from mathematics. Science has inherited from the ancient Greeks its goal of attaining knowledge or understanding rather than mere belief or opinion. The truths of mathematics provided a standard for genuine knowledge. From the time of Galileo to the present day, mathematics has guided the reasoning of basic science. Mathematics is sometimes considered to be a science in itself, but scientists more commonly view it as a fundamental tool—it is the language in which we read the book of nature, as Galileo put it. Just how and why mathematics is so useful in representing the world is a matter of deep philosophical debate, but the history of science has richly demonstrated that it is.

The heart of Greek mathematics and science was geometry. The word *geometry*, meaning "measurement of the earth," is presumed to derive from the response of the ancient Egyptians to practical problems in surveying land and building temples. (The great pyramids exhibit many complex geometric relations.) The Greeks added the method of logical analysis, the very idea of *proving* mathematical theorems by reasoning alone. Euclid's *Elements* (ca. 300 B.C.), a treatise on geometry, is one of the great success stories in world literature. Although more than two thousand years old, this text, which systematized the accumulated knowledge of geometry in ancient Greece, is still used in some schools. (Newton is said to have only once been observed to laugh—when someone asked him what practical use it was to study Euclid.) Einstein, who was deeply impressed as a young boy by the intellectual rigor of geometrical proofs, expressed the central importance of Euclid's work in this comment on the origin of science:

> We reverence ancient Greece as the cradle of western science. Here for the first time the world witnessed the miracle of a logical system which proceeded from step to step with such precision that every one of its deduced propositions was absolutely indubitable—I refer to Euclid's geometry. This admirable triumph of reasoning gave the human intellect the necessary confidence in itself for its subsequent achievements. If Euclid failed to kindle your youthful enthusiasm, then you were not born to be a scientific thinker. (1954, 271)

If one had to choose which single individual has had the most influence on the whole course of science, a good candidate would be ancient Greek philosopher and mathematician Pythagoras (ca. 560–480 B.C.), who introduced the very term *mathematics.* Everybody knows his name from the Pythagorean theorem: The square of the length of the hypotenuse of a right triangle equals the sum of the squares of its sides. A pupil of Thales (624–546 B.C.), who is often cited as the first philosopher/ scientist, Pythagoras settled in southern Italy, where he attracted disciples and formed a secret society. The Pythagorean vision is one of universal harmony that can be interpreted by numbers and their relationships. This vision may have been inspired by an analysis of the pitches of the notes in the musical scale in terms of the arithmetical ratios of the lengths of strings of a lyre. In his *Timeus*, Plato developed a cosmology based on the Pythagorean thesis that numbers and geometric figures are the essence of reality. This was the most influential of Plato's dialogues in Europe during the Middle Ages because it was the only one known (in a Latin translation) before the Renaissance.

The Greek tradition greatly revered intellectual activity, as opposed to manual labor, which was held to be fit only for slaves or craftsmen and thus was beneath the dignity of the lovers of wisdom. One of the Pythagorean maxims expresses a pure, unworldly attitude that Plato impressed on the scholarly tradition for almost two thousand years: "a theorem and a step forward, not a theorem and six pence." Later, Euclid exhibited this attitude that knowledge should be its own reward in his response to a student who asked, "But what advantage shall I get by learning these things?" Euclid told a slave to "give the fellow three pence, since he must needs make profit out of what he learns" (quoted in Van der Waerden 1961, 196).

Plato offered a philosophical basis for this belief that pure theory is of a higher order of value than practical matters by downgrading the whole realm of "opinion" based on perceptual experience. In his dialogue *The Republic*, genuine knowledge is said to concern only the supersensible, eternal Ideas or Forms that are seen with the mind's eye, concepts and images that we innately "remember":

> Those who deal with geometrics . . . although they use visible figures and argue about them . . . are not thinking about these figures but of those things which the figures represent; thus it is the square in itself and the diameter in itself which are the matter of their arguments, not that which they draw; similarly, when they model or draw objects, which may themselves have images in shadows or in water, they use them in turn as images, endeavouring to see those absolute objects which cannot be seen otherwise than by thought. (1961, vi.510)

Perhaps the first great scientist, in something like the modern sense of the word, was Archimedes (287–212 B.C.), who lived a generation after Euclid. His work, which shows a degree of mathematical sophistication not regained until the seventeenth century, was an inspiration to Galileo. Archimedes might also qualify as the first great engineer, except that he didn't have the practical attitude of modern engineers. He was famous for his applications of scientific knowledge to military engines, but according to Plutarch he regarded "as sordid and ignoble the whole trade of engineering, and every sort of art that lends itself to mere use and profit" and "would not deign to leave behind him any commentary or writing on such subjects," which he considered mere "diversions of geometry at play" (1864, 378). The death of Archimedes at the hands of a Roman soldier while making mathematical diagrams in the sand has been seen as symbolizing what mathematician and philosopher Alfred North Whitehead (1861–1947) described as "a world change. . . . The Romans were a great race, but they were cursed by the sterility which waits upon practicality. They were not dreamers enough to arrive at new points of view, which could give more fundamental control over the forces of nature. No Roman lost his life because he was absorbed in the contemplation of a mathematical diagram" (quoted in Turnbull, 1951, 26).

Looking back on the development of science over the last three centuries, we can find some close approximations to the ideal of pure intellectual curiosity. Our first case studies of scientists pursuing knowledge for its own sake focus on two of the leading figures of the scientific revolution that began the modern age.

Kepler: Discovering the Real Motions of the Heavens

Nicolaus Copernicus (1473–1543) began the great change in astronomy with the publication in 1543 of *On the Revolutions of the Heavenly Spheres*, in which he proposed that the earth moves around the sun rather than vice versa. Like Kepler after him, Copernicus was inspired by a love of mathematical simplicity and symmetry as well as a special feeling for the central role of the sun. *Almagest*, by Egyptian astronomer Ptolemy (ca. 190–268 A.D.), had systematized ancient astronomical knowledge into a coherent theory, describing the apparent motions of the heavens by means of a complex system of *epicycles* (circles within circles) while retaining the seemingly obvious observation that the heavens revolve daily about the earth. What distressed Copernicus about the Ptolemaic system were its untidy mathematical features, which he viewed as obnoxious blemishes; he assumed that the true system of the heavens must exhibit an aesthetically satisfying unity and perfection appropriate to its Creator. In fact, his new heliocentric system did not fit the observational data then available any better than its ancient rival, and it only succeeded in reducing, without eliminating, the number of epicycles needed. Copernicus did, however, offer a more unified system (Ptolemy's epicycles had to be devised independently for each planet), and he thus made a significant advance in the mathematical understanding of nature.

Johannes Kepler (1571–1630) is a remarkable figure in the history of science. He was one of the most idealistic searchers for knowledge for knowledge's sake, one of the most enthusiastic and intensely motivated scientists of all time, who burned with desire to understand the motions of the heavenly bodies. Kepler had originally intended to join the clergy, a profession that seemed suited to his frail health—smallpox as a child had left him partially crippled and with impaired vision. However, at the University of Tuebingen in Germany he showed a flair for mathematics and was introduced to the new Copernican theory by a Pythagorean master. In 1594 he began teaching at the University of Graz in Austria. As a professor of astronomy, he spent much of his time with astrology, as was expected in those days. In addition to casting horoscopes, he studied the Greek tradition of astrology, seeking ways to make it more "scientific." He used ideas from astrology to interpret the Bible, leading him to conclude, for example, that the creation of the world occurred in 3992 B.C. In later years Kepler's interest in astrology faded, but he retained a strong strain of mysticism in all his work, a characteristic that made the more skeptical Galileo suspicious of his theories.

Inspired by Pythagorean mysticism merged with a devout Christian faith in an all-wise, all-good, all-powerful Deity, Kepler firmly believed that God's design could be revealed in the austere beauty of mathematics. In his first metaphysical vision of how the heavens should work, he constructed a model for the orbits of the planets based on the five platonic perfect solids, with all faces identical: the tetrahedron (pyramid), the cube, the octahedron (with eight equilateral triangles as faces), the dodecahedron (composed of twelve pentagons), and the icosahedron (composed of twenty equilateral triangles). The spherical shell containing the earth's orbit, for example, was supposed to be circumscribed around the icosahedron in which the sphere of Venus was inscribed. The model worked fairly well for the planets known at the time. Kepler was exuberant at first and commented about his theory in the preface to *Mysterium Cosmographicum:*

> It is attested as true in my deepest soul, and I contemplate its beauty with incredible and ravishing delight. . . . The intense pleasure I have received from his discovery can never be told in words. I regretted no more the time wasted; I tired of no labor; I shunned no toil of reckoning, days and nights spent in calculation, until I could see whether my hypothesis would agree with the orbits of Copernicus, or whether my joy was to vanish into air. (1981)

His joy was to vanish later when he had to reject his platonic solids model because it did not quite fit Tycho Brahe's observations. These were more accurate, despite having been made without the use of a telescope, than what Kepler referred to as "the less precise observations of popular astronomy" of the time. He accepted the refutation but did not give up his Pythagorean faith that mathematical regularities were there to be discovered. He required many theoretical ideas or "visions," much data, and a great deal of sweat to put the data and the theories together. His work exhibits clearly the peculiar psychology required in great science. He wanted to believe beautiful theories and sought aesthetically satisfying models of the universe, but he refused to accept them unless they fit within the limits of accuracy of the best available observations. Kepler exemplifies Bacon's metaphor of the scientist as a bee; unlike the mere experimenters, who like the ant "only collect and use," or the mere reasoners, who "resemble spiders [in that they] make cobwebs out of their own substance," the bee "takes a middle course; it gathers its material from the flowers . . . [and] transforms and digests it by a power of its own." Bacon prophesied that by utilizing "the two faculties, the experimental and the rational, (such as has never yet been made) much may be hoped" (*Works,* vol. 4, 93).

Kepler's most famous contribution to science consists in his three laws that govern the motions of the planets. The first law states that the orbits of the planets, including the earth, are ellipses with the sun at one focus. The second states that equal areas are swept over by a line from the sun to a planet in equal times—so when closer to the sun, a planet moves more quickly. The third law states that the square of the period of revolution of any planet (one year for the earth) is proportional to the cube of its mean distance. These laws show all the planets as functioning in one system, conforming to regularities that Newton was later able to bring under laws governing the whole universe, including the law of universal gravitation.

Kepler was willing to give up an idea that had dominated astronomy for more than two thousand years, that heavenly bodies must move in perfect circles. In contrast, his great predecessor Copernicus had never considered deviating from this belief in circular motion. Galileo, too, never gave up this idea (which is traceable to Pythagoras); his principle of inertia was that the natural motion of unconstrained bodies was circular.

Kepler has a strong claim to the title of founder of physical astronomy, in which, unlike the tradition of Platonism, the aim is to describe and explain the actual motions of real heavenly bodies. What exactly was his achievement? Nineteenth-century American philosopher Charles Peirce saw in his work "the greatest piece of Retroductive Reasoning ever performed" (quoted in Wilson 1972, 92). How then did Kepler come to discover that the planets trace out ellipses rather than circles? The usual account goes as follows: He began by trying to determine the shape of the orbit of Mars—a problem that had caused the greatest difficulty in fitting the Ptolemaic scheme to observations. He plotted Brahe's observations using trigonometry to get from the sightings to the sequence of positions. Then he calculated the various distances of Mars from the sun by triangulation of observed positions. Finally, looking at the plotted data, he saw that it fit an ellipse.

However, a closer look puts Kepler's discovery of his first law in a rather different light. In a carefully documented study, Wilson (1972) argued that Kepler didn't arrive at an elliptical orbit for Mars from the data but rather inferred it by first assuming his second law concerning areas and times. Working this out required 900 pages of horrendous calculations. Kepler's "war on Mars," as he called it, took four years of great dedication and perseverance. He took seriously the eight-minute discrepancy of observations from Copernican theory, which assumes a circular orbit (Brahe's data was estimated to be accurate within 2 minutes of arc). In having to calculate how far Mars traveled in a given time, he faced a problem that was later much simplified by the new mathematical technique of integration introduced in Newton's calculus. Kepler had to do it the hard way, by many tedious computations.

Copernicus had postulated circular orbits for the planets about the sun. He considered the sun to be the proper center of the cosmos, "for who could in this most beautiful temple place this lamp [the sun] in another or better place than that from which at the same time it can illuminate the whole." For Kepler, however, the sun had to be "more than a lamp"; he required it to have a physical function. He believed that the sun by some "virtue" (that is, power or force) caused the motions of the planets, that the planets felt an attraction to the sun that varied according to their distances from it. In the second edition of *Mysterium Cosmographicum,* he remarked: "I used to believe that the cause responsible for the motion of the planets was unquestionably a soul. But when I considered that this moving cause diminishes with distance, and that the sun's light is also attenuated with the distance from the sun, I concluded that this force is something corporeal" (1981).

He wanted to know *why* the planets moved the way they did, what made them tick—or rather, sing, since he believed they actually generated musical tones. (He even used an analogy from music in deriving his third law, a striking example of arriving at correct conclusions by using mistaken assumptions.) He never lost his faith that there was order to be found in the heavens, but he was not content with hypotheses that merely "saved" (in other words, fitted) the appearances. In seeking true physical causes, Kepler marked the transition to modern science more profoundly than Copernicus. The modern scientific revolution began with the heliocentric model, but Kepler added something new, a desire to understand the mechanisms behind the heavenly motions.

Newton: Paragon of the Modern Scientist

Isaac Newton (1643–1727) is at or near the top of everyone's list of great scientists. He developed a theory of mechanics, performed research on optics, and invented calculus—any one of these accomplishments alone would have ensured his place in history. He was virtually deified in his own lifetime. Scottish philosopher David Hume (1711–1775) later declared that he was the "greatest and rarest genius that ever rose" (1864, 481). Nineteenth-century philosopher of science William Whewell stated baldly that Newton's *Principia* (*"Mathematical Principles of Natural Philosophy"*) was "indisputably and incomparably the greatest scientific study ever made" (1857, 136–137). More recently, Stephen Hawking, the present Lucasian professor at Cambridge (the position Newton occupied) has reiterated the common judgment that Newton was "a colossus without parallel in the history of science" (quoted in Fauvel et al. 1988, 27).

Newton fit some of the popular stereotypes of the classic scientist. He was curious in several senses of the word. He could be very absent-minded when concentrating on a problem and is said to have forgotten to eat when engrossed in his work. Like most of the great scientists, he exerted himself to the limit when doing research. In the period between 1665 and 1672, he was reported to have been "in a state of virtually uninterrupted creative exhilaration and exhaustion, studying, pondering so intensely that he took little recreation or sleep." Newton claimed that the secret of his discoveries was to devote intense concentration on whatever problem he was considering: "I keep the subject constantly before me and wait till the first dawnings open little by little in the full light" (Newman 1961, 121). He was very exacting in accounting for observational discrepancies in his experiments. In his work on optics, having noted a deviation of less than one hundredth of an inch in the curvative of a lens he was using, he could not rest until he had found the source of the error. His capacity for taking pains is illustrated by a page in his notes on which he calculated the area under a curve by adding terms of a series expansion to *fifty-five* decimal places.

The famous story of Newton's theory of gravitation being first inspired by the fall of an apple seems to have some truth in it, if interpreted properly. Late in life, Newton was reported to have recalled, "When I was thinking of what pull could hold the moon in its path, the fall of the apple put it into my head that it might be the *same* gravitational pull, suitably diminished by distance as acted on the apple" (More 1934, 44). One reason he did not try at that time (in 1666, at the age of twenty-four) to publish his hypothesis that the force of gravitational attraction follows an inverse square law was that it did not seem to agree precisely with calculated consequences (his error was in using the rough estimate in use among sailors that a degree of latitude corresponded to sixty miles). It was many years later that, with a more accurate measure of latitudes, he corrected his calculations and confirmed his initial hypothesis.

If Copernicus, Kepler, and Galileo set off the great scientific revolution that inaugurated the modern era, Newton essentially completed it—not (as we now know) by having the last word in physics but by determining the way science is done. Newton's resolution of problems of motion set a standard (for better or worse) for what all science should be like—a paradigm that is still influential today. Cosmologist Hermann Bondi contended that "the rails of thinking in physics and astronomy tend to be still very much those that Newton laid down" (Fauvel et al. 1988, 244). He went on to suggest that unfortunately the biological and social sciences have tended to aim for the same kind of solution that Newton so brilliantly achieved for the solar system.

Newton was not humble about his achievement. In Book III of *Principia*, he proclaimed, "I now demonstrate the frame of the System of the World" (1962, 397), and he proceeded to do just that. His underlying goal was even more ambitious: to solve the mysteries of the universe, reading the "riddle of the Godhead" by deciphering the clues God had provided the human race. As scholars continue to examine his voluminous unpublished writings in theology, alchemy, and neoplatonic magic, it turns out that Newton was a mystic as well as the coolly rational scientist he is generally pictured as having been. An extreme expression of this more common image is the remark of nineteenth-century English writer Charles Lamb, who saw Newton as "a fellow who believed nothing unless it was as clear as the three sides of a triangle" (quoted in Fauvel et al. 1988, 6).

In his publications, Newton separated his metaphysical investigations from his natural philosophy, but they are not unconnected with his motivation, his overall goals. He had the Pythagorean faith that beneath the appearances of our sense experience lies a deeper reality that can be revealed by mathematics. His much-discussed claim that he framed (or "feigned") no hypotheses (*"hypotheses non fingo"*) might be interpreted as excluding *from natural philosophy*, that is, from science, any speculation about ultimate causes, especially of gravitational attraction. He has presented us with one of the great paradigms for all time of the mathematical understanding of nature. In 5.1 we shall look at the less attractive side of Newton the scientist, administrator, and public man.

Einstein: Fascination with the Universe

Albert Einstein (1879–1955) was by all accounts the greatest physicist since Newton. Even as a schoolboy, Albert had formed his project in life of seeking to understand the fundamental laws of the physical universe. At the age of five, he was entranced by the curious behavior of a compass needle. At age twelve, he became fascinated by geometry upon encountering the proof of Pythagoras's theorem. His education was patchy—he seems to have been an introspective boy, determined to go his own way. At the Swiss Federal Institute of Technology in Zurich, he skipped many lectures, teaching himself theoretical physics by studying the texts on his own. Needing to earn a living to support his family, he unsuccessfully tried teaching and then found employment in the Swiss Patent Office. Without any academic status, lacking contact with any of the foremost physicists of the time, and with the demands of his job examining patent applications, one would not expect him to produce any great scientific results. Yet in 1905, while still working at the patent office, Einstein published three momentous papers in volume 17 of *Annalen der Physik* (a copy of which is today worth several thousand dollars). One of these papers, a discussion of Brownian motion, supplied the definitive arguments for the physical existence of molecules; another, an examination of the photoelectric effect, contained a pioneer application of the idea of quanta of energy; but it was his brief paper entitled "The Electrodynamics of Moving Bodies," which was the first exposition of his theory of relativity, that revolutionized the very concepts of space and time.

To say he was a genius is merely to label the phenomenon but not to explain it. A couple of years before his death, Einstein himself denied that he possessed any special talents, though he went on to cite some of the qualities required for great scientific work: "Curiosity, obsession, and dogged endurance, combined with self-criticism, have brought me to my ideas" (1954). His revolutionary papers could not have been conceived without his having familiarized himself with the fundamentals of what was already known in physics, followed up by tireless effort to work out the consequences of his hypotheses. Obviously he did have genius, in the form of a kind of physical "intuition" that inspired his novel ideas, but his ideas would not have been fruitful if not allied with a determination to formulate them precisely in such a way that they could be rigorously tested. Nobody would undertake such mental labor, especially while employed full-time and burdened with the responsibilities of family life, without a burning desire to understand aspects of the physical universe previously hidden from human understanding.

Einstein expressed this most basic passion motivating his life in these words: "I'm not much with people, and I'm not a family man. I want my peace. I want to know how God created this world. I'm not interested in this or that phenomenon, in the spectrum of this or that element. I want to know His thoughts; the rest are details" (quoted in Clark 1971, 18–19). The reference to God here does not indicate a belief in a personal deity but rather the idea of a rationally ordered nature. Einstein was wont to express his approach to science in such quasi-religious terms. His famous remark, in response to a threatened complication to his theory of relativity in 1921, that "the good Lord is subtle but not malevolent" (Clark 1971, 390)—or in Einstein's own translation to Derek Price, "God is slick, but he ain't mean" (Mackay 1991, 52)—expresses his conception of nature's secrets as open to human reason, if only after much imagination and toil. The drive to understand the most general, fundamental aspects of nature never left him; it motivated him just as strongly in his long and largely fruitless theoretical struggles in later life, as we shall see. His statement that he was not a family man is revealing. Clearly his passion for physics came before everything else in his life. He married Mileva, a fellow student at Zurich, in 1903. It has been suggested from correspondence recently made public that she had some influence on Einstein's early scientific ideas. Mileva and he had two sons, but they separated after eleven years. Einstein then married his cousin Elsa, who was prepared to take on the role of looking after all his domestic and practical affairs, protecting the great thinker from the world.

Einstein's papers of 1905 brought him into the center of the physics community. He started to attend the important congresses and to impress his fellow physicists. Soon he was out of the patent office and into a succession of academic positions—in Berne, Zurich, Prague, Zurich once again, and then finally at the newly established Kaiser Wilhelm Institute at Berlin in 1914. The prospect of the intellectual stimulation made possible by the presence of other theoretical physicists there seems to have been what overcame his reluctance to return to Germany, a country whose militaristic ethos he had begun to dislike even as a schoolboy. The move was indeed fruitful scientifically, for in Berlin (during World War I) Einstein was able to complete his supreme intellectual achievement: the general theory of relativity. The observational confirmation of his relativistic theory's superiority to Newtonian mechanics (by the bending of light near the sun, observed during a solar eclipse soon after the war) earned him enduring international fame. But he neither sought nor enjoyed this. Reflecting later on his scientific career, he commented in a speech before the American Academy of Sciences in 1921: "When a man after long years of searching chances upon a thought which discloses something of the beauty of this mysterious universe, he should not therefore be personally celebrated. He is already sufficiently paid by his experience of seeking and finding."

Nor did he seek personal profit or power within the scientific community. His supreme reputation led to innumerable offers of academic positions in universities in many countries of the world, and he could have gone almost anywhere he liked, on his own terms. Though he hesitated many times, his primary criterion was not how much money was involved (by this time he had enough to support his family, and his own needs were minimal), but which place could offer the best intellectual environment for his scientific work. This seems to be what kept him in Berlin through the 1920s, despite becoming a target of increasing anti-Jewish sentiment.

When Hitler came to power in 1933, Einstein was on a visit to the United States, and there was no way he would return to Germany. He became one of many prominent scientists to take refuge in the United States, where he found a permanent home at the Institute for Advanced Study at Princeton University. In his remaining years, he continued to try to develop a unified field theory that would bring together the basic concepts and laws of physics into a single comprehensive statement—but without success. Remaining true to his deeply felt intuition about the rationality of nature, he kept up a losing struggle to show that the notorious indeterminism in quantum mechanics was only apparent, that it covered up an underlying determinism. Most physicists were by then agreed that the theory was irreducibly probabilistic, but Einstein could never bring himself to accept the Uncertainty Principle of Werner Heisenberg and Niels Bohr; he maintained to the end that "God does not play dice with the universe" (written in a letter to Max Born in 1926 and quoted in Clark 1971, 340).

At a time of life when many scientists rest on their laurels (and nobody had more right to do so than Einstein), he continued to his dying day to work on the most abstruse and difficult problems of physics and cosmology. He said that these were projects suitable only for someone like himself who had already made his name, having indeed attained such fame that subsequent failures could hardly detract from it. He claimed to be doing physics a service in tackling problems that nobody else would face. Albert Einstein was a magnificent combination of physical intuition, determination to struggle with the hardest problems, and nobility of character and sentiment, even if combined with a certain remoteness from the ordinary human world. In 5.3 we shall look briefly at his attitude toward public fame.

Gibbs: Revered by a Few, Unknown to Many

If you were to ask a physical chemist who the greatest American-born scientist was, he or she would probably name Josiah Willard Gibbs (1839–1903). Max Planck, one of the fathers of quantum physics (see 8.1), considered Gibbs to be one of the greatest theoretical physicists. Yet many people have never heard of him, and in one extensive anthology of American scientists he is not even mentioned. Gibbs did important work in thermodynamics and laid the foundations of physical chemistry, which later contributed to many industrial applications. His work gained him honors from the American Academy of Science, the Rumford Medal from the Royal Institution, and the Copley Medal from the Royal Society. W. P. D. Wightman, in a classic history of science, stressed that "Gibbs must be numbered among the greatest and most influential of human intellects" (1950, 296). At least among a select few, then, he has attained a very high scientific reputation.

There is not much to relate about his life. He lived in New Haven, Connecticut, except for three years spent as a young man studying in Europe. An early interest in engineering led him to invent a hydraulic turbine and an improved train brake. He received the first U.S. doctorate in physics from Yale University, where he went on to teach as professor of mathematical physics. He never married and lived simply in a room at the home of his sister. He was singularly unambitious for fame. The story is told that when some European physicists, who recognized the greatness of his work, came to the United States to visit him, the president of Yale thought they were looking for Wolcott Gibbs, a then prominent chemist.

One reason he never attained public fame is that his published work was highly theoretical and was presented in a mathematical formulation that many found hard to understand. Even his now much-used phase rule (originally described by five pages of mathematical equations in a paper written in 1875) was not appreciated at the time. (One early application, however, was the attempt by Henry Adams in 1909 to apply Gibbs's phase rule to the dynamics of progress in history!) The phase rule tells how to determine the concentrations of various substances to get a desired mixture with the components in equilibrium, in which state they don't separate out. During World War I, German chemist Fritz

Haber (see 6.2) made a very practical application of Gibbs's work in perfecting a synthesis of ammonia gas, which served as a source for nitrates to make explosives and fertilizer.

In striking contrast to the way science is conducted today, a century after his time, Gibbs had no laboratory and no graduate assistants. In fact, for the first nine years that he taught at Yale, he was not paid any salary at all. Since he had an inheritance sufficient to allow him a modest living, a salary was thought unnecessary. Only after Johns Hopkins University offered him a position did Yale give him a modest stipend. His courses were difficult, but he was considered a good teacher by the really dedicated students. Gibbs was content to spend his life absorbed in his scientific work. The daughter of his colleague and close friend H. A. Newton once described him as "the happiest man she ever knew" (Wheeler 1962, 176). But exceptional scientific merit does not guarantee worldly rewards, in terms of money, power, or fame.

4.2 Theoretical Insight and Experimental Skill

So far we have looked at scientists who were primarily theoreticians. Although Newton did some experiments (with prisms, on the refraction of light), Kepler and Einstein did not. The solar system and the space-time structure of the entire universe do not lend themselves to human experimentation, of course, but the truth is that Kepler and Einstein were not great scientific observers; they depended on the skill of others for the verification of their theories. Like Newton, they were supreme theoretical physicists. We will now examine some cases where theory and experiment, intellectual curiosity and practical skill, were closely bound together.

Mendel's Lonely Breakthrough in Genetics

A good example of the ideal of the pure (and in this case, literally monastic) scientist is Austrian monk Gregor Johann Mendel (1822–1884). He has become famous—but only after his death—as the father of the twentieth-century science of genetics. In 1900 his work was rediscovered, and eventually his ideas were integrated with Darwin's theory of natural selection to form evolutionary theory as it's commonly understood today. Mendel's pioneering experiments and theories on the basis of heredity also laid the foundation for the investigation in the 1950s of the biochemical basis of genetics in the structure of the DNA molecule (see 5.1).

Born into a poor peasant family in Moravia (then part of the Austro-Hungarian Empire, now part of the Czech Republic), Mendel had to struggle to get a proper education. He entered the Augustinian monastery at Brunn at the age of twenty-one. This was a center of learning in mathematics, music, and natural science at the time, and Mendel found opportunity there to pursue his scientific interests. However, he twice failed a qualifying exam for a high school teaching license (he passed the part on physical sciences, but ironically enough, he failed in natural history!). From early on, he showed a fondness for growing things and a great curiosity about the mysteries of life. He asked, "Must we throw up our hands and say that nature in this basic life process is completely irrational or incomprehensible?" Before making his discoveries about heredity, he recorded his conviction that "there must be definite laws of heredity" (Sootin 1959, 116).

Animal breeding was forbidden by Mendel's religious superiors because it seemed to them to involve playing with sex, but there was a tradition of plant breeding in the monastery garden; the reproduction of plants was apparently considered innocent enough! A modest garden strip 120 by 20 feet in size was his laboratory. His simple equipment, which consisted of forceps, a camel hair brush, and small paper bags to confine pollen, contrasts sharply with the expensive and technically complex experimentation of today. Mendel's meticulous research required great perseverance. He toiled for seven years, cross-breeding some 30,000 individual pea plants.

His published papers show that he had thought deeply beforehand about what he was trying to prove in his painstaking experiments. He had formulated the hypothesis that the various characteristics of sexually reproducing organisms were passed on from one generation to the next by random combinations of certain distinct factors (later identified as genes) supplied by the parents of each individual. Because of his mathematical training he was able to formulate an algebraic model of all the possible combinations of genes, which enhanced clear understanding. It also allowed testable quantitative predictions for the proportions of various characteristics in successive generations. By combining all this with an innovative application of simple statistics, Mendel pioneered the application of mathematics to biology, which has greatly advanced biological theory in the twentieth century.

Though interested in the general question of heredity of living organisms, Mendel shrewdly set out to test it exhaustively on one particular species of plant. He choose the pea because it exhibited easily distinguished variations in size, color, and shape of seeds. Had he started with other plants, it is likely that the complexities of inheritance, which were only much later explained, would have led him into confusion. Using statistical analysis of large numbers of various "crossings" (breedings of two individuals with different characteristics—now standard methodology, but then quite new), Mendel achieved decisive verification of his hypotheses. In retrospect, he can perhaps be excused for having presented his data in somewhat neater ratios than are likely to have actually occurred. Statistician R. A. Fisher (1890–1962) argued that "the data of most, if not all of the experiments have been falsified so as to agree closely with Mendel's expectations" (Fisher 1936, 115). There are always temptations for scientists to interpret raw statistical data in the light of their theoretical hypotheses.

In some ways, Mendel's achievement might not seem so impressive a revolution in understanding. Many before him had done cross-breeding experiments, and some had speculated as to how heredity works. But it was he who put together the theoretical foundations for genetics by deriving the mathematical relationships of the patterns of inheritance in a simple model of hypothesized causal factors (the genes). He thus followed the approach that had already been so successful in physics and astronomy in seeking to understand how nature works; namely, to look for mathematical patterns, hypothesize causes, and then test observational predictions of the model.

Mendel presented his results in a lecture to the Brunn Society for Study of Natural Science. His paper "Experiments on Plant-Hybridization" did not excite much enthusiasm in the audience, and neither did its publication in the *Proceedings of the Moravian Society* attract much attention. He sent copies to various scientific societies and universities—and to Darwin, whose works he had studied carefully. If he read Mendel's paper at all, Darwin failed to appreciate its importance for evolutionary theory. Mendel was to publish only one more scientific paper, but he tried to interest Karl Naegeli (1817–1891), a prominent Swiss botanist of the time, who also failed to realize the fundamental significance of his work. In subsequent years, he made further experiments with other species of plants and with bees, but he encountered technical difficulties that could not be overcome. He might have gotten the attention of the scientific community if he had been in closer contact with other biologists, but such interchange is just what he lacked. So he died unrecognized as the brilliant scientist he was.

As we said before, Mendel was a scientist who lived a literally, rather than metaphorically, monastic existence. The monastery served for him rather like a university does for today's scientists, giving financial security (at a modest level), experimental facilities, and some time to pursue research. Marriage was not an option, of course (in fact, many of the early great scientists were single; perhaps some ambitious scientists today, facing demands on their time by their families, might occasionally envy Mendel his freedom in that respect!). However, after being elected abbot of the monastery in 1868, his administrative duties restricted his scientific research. He became something of a man of affairs, as government counselor and bank chairman. Though generally described as a cheerful, kindly, and modest person, he became involved in disputes concerning tax assessment on the monasteries.

Did Mendel do his research simply for the satisfaction of finding out the truth (or for the glory of God?), without any desire for recognition or fame? He must have derived considerable fulfillment from doing his work, enough to see him through such a long, tedious period of organizing and recording pea plants. But his letters to Naegeli show him wanting to communicate his scientific ideas and receive expert comment on them. He expressed hope that others would take his theories seriously enough to test them on other species. However indifferent he may have been to personal fame, he did recognize the inherently collaborative nature of science, and he must surely have been deeply disappointed by how little recognition his beautiful work in genetics received.

Madame Curie's Determined Pursuit of Radioactivity

Let us now consider the fascinating story of Marie Sklodowska, better known as Madame Curie (1867–1934), the first woman to become a top-ranking scientist and the first person to win a Nobel Prize twice. She provides another example of selfless devotion to pure scientific research, fueled by a steely determination that sustained her through a hard life. Marie was born in the portion of Poland repressively ruled by Russia throughout the nineteenth century. Fascinated by her father's physics apparatus, she felt drawn to science in her school days but had no hope of a scientific education in the Warsaw of those times. She did attend some classes at the informal "Floating University," which operated clandestinely despite the restrictions imposed by the Russian rule of that part of Poland, but her only prospect for scientific training was to save enough money to go abroad. This she managed to accomplish by persuading her elder sister Bronya to go to medical school in Paris, financed by Marie's savings from working as a governess. The agreement was that when Bronya started earning money, she would pay for Marie to come to France to study. So for several years, Marie had the lowly job of watching over the children of aristocratic families in Poland, using her spare time to pore over whatever scientific books she could find.

When at last her chance came in 1891 to go to Paris and enroll in the Sorbonne, she seized it with enthusiasm. Studying with passionate devotion, stretching her meager resources by living in unheated garrets, hardly bothering to eat, sometimes fainting from hunger and exhaustion—this era of her life may seem like the operatic cliché of the bohemian student. But Marie Sklodowska was no caricature; she was a remarkable young Polish woman with no time for relaxation or love affairs, determined to learn all she could about chemistry and physics and possessed of an ambition to contribute to scientific knowledge.

After completing her bachelor's degree in physics with highest honors in 1893 and receiving a master's degree in mathematics in 1894, she met French physicist Pierre Curie, whose dedication to science was as austere and selfless as her own. Their common intellectual devotion led to friendship, which rapidly became love, and they were married in 1895. But Marie allowed neither motherhood (she had two daughters, one of whom also won a Nobel Prize) nor uncertain health to interrupt her scientific career. Nurses were engaged for the children, and the daily work at the laboratory resumed.

She chose as the topic for her doctoral research the mysterious radiation, emitted by salts of uranium, recently discovered by Henri Becquerel (1852–1908). What was the nature of such radiation? What was its cause? Marie found that the strength of the radiation was proportional to the amount of uranium under examination and, moreover, that it was independent of light, temperature, or chemical combination; she concluded that radiation must be an atomic phenomenon. If so, could it not appear in other substances? She tested all the known elements and found that thorium emitted similar radiation, which she named *radioactivity*. She then proceeded to examine all available ores and eventually found a strong source of radiation in certain samples of pitchblende that was much more powerful than uranium. She first thought this had to be due to experimental error, but after measuring and remeasuring, she suspected that there might be a hitherto unknown but strongly radioactive element present in pitchblende in very small quantities. Her research was reported in a paper published in 1898, but to convince the scientific world, it was necessary to prepare a pure sample of the hypothetical new substance.

Pierre was doing research on crystals at the time, having earlier discovered piezoelectricity (which results when pressure is applied to certain crystals). But he put aside his own research to join in his wife's work. Their marriage became a collaboration in a single research project, which continued as long as they both lived. The Curies did not have anything like the facilities that scientists expect nowadays. Because of his unworldliness and lack of professional influence, Pierre had no laboratory he could call his own except a damp, cold shed. (It is said that once when Pierre was told that the French government was considering him for inclusion in the Legion of Honor, he responded that he would prefer a decent laboratory!) In this shabby workshop, amid noxious fumes, they took on the long and arduous process of isolating two new elements: first polonium (named after Marie's native land) and finally radium, as the powerfully radioactive substance was called.

For nearly four years, Marie worked daily on the repetitive processes of purifying tons of pitchblende under conditions later realized to involve exposure to dangerous levels of radioactivity. The raw materials and equipment had to be paid for at the Curies' own expense. Besides solving the intellectual problem of devising methods to extract the elements from the raw ore, they had to contend with the physical drudgery of carrying them out. Marie chose to concentrate on the latter, as if the traditional marital division of labor gave her the task of "cookery." It seems that Pierre would gladly have postponed the project, but Marie, having started the job, was determined to finish it despite the inadequacy of their equipment. Her unrelenting commitment carried them through to a successful conclusion. A stream of publications about radioactivity followed, and the couple suddenly found themselves at the center of world scientific attention. Honors followed—they received the Royal Society's Davy Medal and, with Becquerel, were awarded the Nobel Prize for physics in 1903. The money from the latter rescued the Curies from the edge of poverty.

Pierre died in a street accident in 1906 (it seems likely that deleterious effects of their experimental work may have reduced his alertness). Here was another occasion on which only Marie's intense commitment to science kept her going. She was offered the Chair of Physics at the Sorbonne, which Pierre had held. Steeling herself to resume her scientific vocation, she faced a curious crowd eager to hear the first woman to teach at that prestigious institution, and she deliberately took up the subject at the very point where her late husband had left it in his last lecture. Further research led to her second Nobel Prize in chemistry in 1911.

By the 1920s, her fame was worldwide, especially as the first woman scientist of note. But honors and adulation did not affect her. She could only be persuaded to undertake tours of the United States, for example, if there would be some gain for her research center, the Curie Institute, as in 1921 when she was presented with a gram of radium by President Warren Harding as a gift from American women. The words she wrote concerning the need for scientific scholarships apply to her own career:

> What is society's interest? Should it not favor the development of scientific vocations? Is it, then, rich enough to sacrifice those which are offered? I believe, rather, that the collection of aptitudes required for a genuine scientific vocation is an infinitely precious and delicate thing, a rare treasure

which it is criminal and absurd to lose, and over which we must watch with solicitude so as to give it every chance of fruition. (Curie 1937, 340)

A rare treasure indeed was the "collection of aptitudes" that this extraordinary woman presented to the world: resolute determination to pursue her scientific curiosity, a fixity of purpose undaunted by the difficulties of getting an education, of poverty, of arduous labor, and of motherhood or widowhood.

Einstein gave a moving portrait of what made Madame Curie so outstanding:

I came to admire her human grandeur to an ever growing degree. Her strength, her purity of will, her austerity toward herself, her objectivity, her incorruptible judgment—all these were of a kind seldom found joined in a single individual. She felt herself at every moment to be a servant of society and her profound modesty never left any room for complacency. . . . Once she had recognized a certain way as the right one, she pursued it without compromise and with extreme tenacity. (1954, 77)

Raman's Research on Colors and Sounds in India

Our third example of theoretical curiosity and experimental skill is Indian physicist Sir Chandrasekhara Venkata Raman (1888–1970), whose work on the scattering of light earned him the Nobel Prize for physics in 1930. This was one of the first cases of someone from outside Europe and the United States making an internationally recognized contribution to science.

Raman graduated in physics with highest honors from the University of Madras, but ill health made it impossible for him to travel to Britain for postgraduate study. The route to a scientific career being apparently closed to him, he entered the Indian civil service in the audits and accounts division. He spent ten years in this well-rewarded profession, taking on increasing responsibilities. But during those years he somehow found the time to conduct research on vibrations, sounds, and the theory of musical instruments at the Laboratory of the Indian Association for the Cultivation of Science. Physics must have been the strongest love of his life. His published research began to earn him a reputation as an important scientist. When the University of Calcutta offered him its Chair of Physics, Raman had no hesitation in accepting, although he was thereby giving up a more lucrative career in the civil service—another case of scientific curiosity outweighing more worldly motives.

Raman's further research on acoustics led to international recognition, and in 1921 he represented the University of Calcutta at a congress in Oxford and lectured to the Royal Society in London. Impressed by the opalescent blue of the Mediterranean Sea on the voyage home, he looked for an explanation of such color effects. He was able to show that light is scattered in a certain distinctive way by water molecules. The work for which he was awarded the Nobel Prize was his discovery of this new type of secondary radiation in light scattered by water and air. The shift of some of the incident light toward longer wavelengths, due to energy absorbed by the scattering molecules, is known as "the Raman effect." Raman spectroscopy has become a widely used technique for investigating materials according to the light they scatter. Raman spectra provide information both for fundamental research on molecular structure and for practical problems of chemical analysis that arise, for example, in the petroleum industry.

He was a "pure" scientist in that he applied himself to topics that had (at the time) no commercial, military, or medical relevance. The focus of his interests seems to have been influenced by Indian culture, with its special appreciation of the beautiful sounds of musical instruments and the bright colors of flowers, gems, and the sea. He was the first to study systematically the color of flowers with a spectroscope. In his later years he became interested in color perception and wrote a book presenting his new theory of the subject.

Raman had great influence in stimulating scientific research in India; indeed, he exercised a dominating kind of professional power in his country. He has been described as proud to the point of arrogance, but he was an excellent teacher who trained hundreds of students. He once remarked that "the principal function of the older generation of scientific men is to discover talent and genius in the younger generation" (Gillispie 1976, vol. 11, 267). He founded the *Indian Journal of Physics* and the Indian Academy of Sciences and also had a hand in building up most of the national scientific research organizations. India has since produced many distinguished scientists.

4.3 Enjoyable Tinkering

Scientific research is not commonly thought of as an especially hedonistic pursuit, but some scientists actually enjoy the process of experimentation in the laboratory, perhaps even if no definite results are achieved. Indeed, experimentalists often take pride in the skillful execution of their craft. These days, we tend to think of scientific research as involving high-tech equipment such as ultracentrifuges or electron microscopes, multimillion-dollar telescopes or massive particle accelerators, with teams of specialist technicians to operate them. But until this century, many scientists, in the tradition of Galileo and Newton, made their own apparatus out of "string and sealing wax," as Lewis Carroll put it.

Pride in experimental craft was vigorously expressed by English physicist P. M. S. Blackett (1897–1974) in an essay entitled *The Craft of Experimental Physics,* written in 1933. He wrote that a scientist must be "enough of a theorist to know what experiments are worth doing, and enough of a craftsman to be able to do them" (quoted in Crowther 1941, 514). A scientist must be able to blow glass, turn metal, do carpentry, take photographs, wire electric circuits, and be a master of gadgets generally—and has to do such things most of the working day. Connecting this with what he supposed to be a national penchant for practical hobbies, Blackett ventured the suggestion that English experimental physics "has derived strength from the social tradition and moral principles which led the growing middle class to spend the leisure of its prosperity in the home rather than the cafe" (514)—a severe put-down for those unreliable continentals who wasted their time discussing ideas in cafes instead of getting on with real tasks in their garages and sheds! Blackett went on to write: "It is the intimate relation between the activities of hand and mind which gives to the craft of the experimenter its peculiar charm. . . . His legitimate field of activity ranges from carpentering to mechanics; it is his job both to make and to think and he can divide his time as he sees fit between both these pleasurable occupations" (515). Blackett himself was a dogged experimenter who took some 20,000 photographs of almost half a million alpha particle tracks in a cloud chamber. He managed to pick out eight tracks that showed collisions of an alpha particle and a nitrogen molecule, which marked the first observation of a nuclear reaction.

Scientist and statesman Warren Weaver has described an early sign that he was a budding experimental scientist; at age seven it was "perfectly clear to [him] that taking things apart and finding out how they are constructed and how they work was exciting, stimulating, and tremendous fun" (Love and Childers 1965, 267). We will now look at two examples of scientists for whom the pleasure of experimenting seems to have been almost as strong a motivation as curiosity about the results.

Priestley Amusing Himself with Gases

Joseph Priestley (1733–1804) was a poor English clergyman (a Unitarian, dissenting from the dominant Anglicanism) who did important work in the development of chemistry. An amateur scientist who loved to perform experiments, he had read books on chemistry and felt the urge to conduct his own research on gases, which was greatly helped by his living next door to a brewery. As he wrote in his autobiography:

I, at first, amused myself with making experiments on the fixed air which I found ready-made in the process of fermentation. . . . When I began these experiments I knew very little of chemistry. . . . But I have often thought that, upon the whole, this circumstance was no disadvantage to me; as, in this situation, I was led to devise an apparatus and processes of my own, adapted to my peculiar views; whereas, if I had been previously accustomed to the usual chemical processes, I should not have so easily thought of any other, and without new modes of operation I should hardly have discovered anything materially new. (quoted in Huxley 1964, 13)

He observed that a colorless gas given off by fermentation in the brewery extinguished burning chips he suspended over the vats. The gas was, in fact, what we now know as carbon dioxide. He found that it could be used to make an "exceedingly pleasant sparkling water." Priestley can thus be credited with discovering soda water! (He received the Copley Medal from the Royal Society for this.) In his homemade lab, he went on to isolate and study a variety of gases, including ammonia. But he is best known as one of those who contributed to the discovery of oxygen.

The question "Who discovered oxygen?" poses interesting historical and conceptual issues, as Kuhn (1962) has emphasized. Priestley is sometimes given the credit, since he was one of the first to collect a relatively pure sample of oxygen gas by heating mercury oxide. Happening to have a candle at hand, he tried inserting it in the flask just to see what would happen. To his surprise, it burned brightly. He dubbed the gas "dephlogisticated air," a description suggested by the phlogiston theory of heat, which was commonly accepted at the time. Phlogiston was thought to be a substance given off when substances burned, and plants were thought to remove phlogiston from the air. Phlogiston was also supposed to possess a mysterious property called negative weight, which explained the measurable gain in weight of things that had burned. But we now understand combustion as oxidation—the addition of oxygen, not the subtraction of phlogiston, and we also know that plants give off oxygen. Everyone had of course been breathing oxygen, but nobody distinguished it as a component of air. Priestley isolated a sample of what we now know to be oxygen and noted some its properties, but he viewed it through the eyes of a mistaken theory. Priestley's French contemporary Antoine-Laurent Lavoisier (see 7.1.2) has a stronger claim as the real discoverer of oxygen (which he named as such), since he was the first to recognize that the gas was a chemical element as now understood in standard chemistry.

Because of his support for the French revolution, Priestley aroused suspicion, even hatred, in the conservative England of his time, and a mob yelling "No philosophers! Church and King Forever!" burned his chapel and home in 1791. He moved to London but was shunned by members of the Royal Society because of his radical political views. He finally emigrated to the United States in 1794. Benjamin Franklin, who had met Priestley earlier during a visit to England, welcomed him and offered to get him a professorship at the University of Pennsylvania. But Priestley preferred to continue his private writing and research. While overseas, he produced carbon monoxide by passing steam over burning charcoal (this "inflammable air" was, however, confused at the time with hydrogen). He also made nitrous oxide, which was later used as an anaesthetic and called "laughing gas." But he did not call these gases by their modern names, for he never accepted the revolution in chemical theory and nomenclature that Lavoisier initiated. Priestley's skill and pleasure was apparently stronger in experimenting than theorizing.

Fleming Playing Around with Germs and Molds

Sir Alexander Fleming (1881–1955) attained worldwide fame, though perhaps not entirely deserved, as the discoverer of penicillin, the first effective antibiotic. The growth of the crucial mold on one of his laboratory plates was a matter of extraordinary luck, and the antibiotic potential of penicillin was developed not by him but by Howard Florey's Oxford team some twelve years later. However, without Fleming's experiments and sharp observations, the potential of the germ-killing mold might never have been noticed.

Fleming worked in St. Mary's Hospital in London for his entire professional life, except for the years of World War I, which he spent trying (with little success) to treat horrible battle wounds in a makeshift hospital in France. His war experience showed him the need for an antiseptic that would kill infectious bacteria without also destroying the body's defensive white blood cells.

Back in St. Mary's after the war, he made the first of his two important discoveries. He had developed the habit, unlike his tidier colleagues, of keeping forty or more living cultures going on his bench, which he would watch to see if anything unusual had grown. Once, when suffering from a cold, he cultured some mucus from his nose. He observed that after two weeks, the mucus seemed to be killing the bacteria in its vicinity. Finding that saliva, tears, and the whites of eggs have a similar effect, he inferred that some antibacterial agent was present in the fluids. Fleming reported his work to the Royal Society in a paper entitled "On a Remarkable Bacteriolytic Element Found in Tissues and Secretions." It turned out, however, that only certain rare bacteria are affected, so his discovery of the germ-killing substance, later isolated and called *lysozyme*, had no therapeutic application.

In 1928, Fleming made the discovery that was eventually to raise him to dizzy heights of celebrity. He had continued to search for a germ killer, and one day he noticed a clear patch in a culture of infectious staphylococcus bacteria. He inferred that the bacteria had been killed by some contamination, which he later traced to the mold *Penicillium notatum.* This mold secretes a liquid that Fleming named *penicillin.* He published an account of his observations, but the scientific world paid little attention.

Fleming's discovery depended in fact on a remarkable set of coincidences. Recent scientific detective work by Hare (1970) has suggested that this breakthrough involved even more luck than usually thought. According to Hare's analysis, the following combination of circumstances was needed for the clear patch to have developed in Fleming's culture:

1. Fleming inoculated a plate with staphylococci, and the plate happened to get contaminated when a spore of a very rare strain of a common mold floated into his lab, probably from a lab below, where another researcher was studying tropical molds. (Fleming was a gregarious character who liked to leave his door open so that people would drop in.)

2. For some reason, Fleming did not warm the plate as usual.

3. He left the plate on his bench while he went on holiday.

4. During that time, cold weather was followed by a warm spell, allowing first the penicillin-producing mold and then the bacteria to grow. (It seems penicillin is only effective against actively dividing cells.)

5. When Fleming returned, he cast an eye over the plate, saw nothing unusual, and discarded it, but by chance the plate did not get washed.

6. Someone popped in for a chat, and Fleming, as was his wont, proceeded to show off the plates lying on his bench.

7. He happened to pick up the plate in question and noticed something very interesting—germs had been killed in the region near the mold.

This may make it sound as if no credit at all is due to this eccentric Scotsman with filthy laboratory habits, who allowed his nose to drip into his experiments, left unwashed dishes all over the place, and kept no proper records of what he was doing! However, Fleming's lab notes show him to have been really quite methodical. He was aware of the subtlety of the mix of factors involved in the growth of bacteria and was constantly on the lookout for any departures from the norm. If he "played around" by allowing a wide variety of cultures to grow on his plates, this was deliberate policy—to see what might happen. The effects of that stray patch of rare mold would probably not have been noticed by a less expert experimentalist and keen observer. Fleming is a good illustration of the "principle of limited sloppiness" in experimentation first enunciated by the physicist-turned-geneticist Max Delbruck (1906–1981): Be sloppy enough that unexpected things show up but not so sloppy that you can't analyze what happened and get the phenomena to repeat (cited in Root-Bernstein 1989, 412). But now we must ask this question: Why, for the next twelve years (1928–1940), did Fleming not pursue the antibiotic potential of penicillin?

There were technical difficulties in getting pure samples of the stuff in any quantity, but he could have gotten by with limited quantities and performed therapeutic trials on small animals such as mice. If he had, he might have succeeded in persuading biochemists to purify larger samples of penicillin. However, Fleming had some reason to think that it would be useless as an antibiotic or even as a local antiseptic, since it disappears from the bloodstream in about half an hour. So he abandoned the project and turned to other research. It was left to others to develop penicillin as the famous antibiotic that has saved so many lives, a story we shall continue in 5.3.

Many find it strange that Fleming should go down in history as a hero of science, since he seems to have lacked the characteristics of industriousness and imagination usually associated with those who make great scientific discoveries. He was a relaxed character who would do his six hours a day in the lab and then repair to his convivial Chelsea club for the evenings; his weekends were spent at at his country retreat. He was, however, a highly competent bacteriologist who literally made an art of his experimental technique and delighted in showing off the "germ pictures" that he produced by subtly staining colors into the cultures grown on his laboratory plates. His talents lay more in experiment and observation than in theory or practical application. For Fleming, the pleasures of doing laboratory work seem to have been at least as strong as theoretical curiosity or the desire to be medically useful; he was not driven to pursue penicillin's potential very resolutely. In the assessment of his biographer Gwyn Macfarlane, "Fleming's level was indeed play. . . . 'I play with microbes'—his often repeated description of his work—was literally true. Most of his research was a game to him"

(1984, 263). The moral is that many different kinds of talent—and sometimes a large slice of luck—can contribute to scientific success.

Despite the great amount of good fortune and serendipity required for Fleming's discovery of penicillin, the idea of using molds to kill disease-causing bacteria had been anticipated. English surgeon Joseph Lister (1827–1912), who introduced antiseptic surgery, noted in 1871 that certain molds prevented the growth of bacteria. He made some experiments, but they didn't come to anything. Something closer to a real anticipation of Fleming's discovery was the work of French medical student Ernest Duchesne. He had been reading Pasteur and became curious about a "struggle for existence" among microorganisms. He noticed that when a single spore of mold (which just happened to be *Penicillium glaucan*) fell on a damp piece of bread or other rotting food, it would grow but sometimes left some patches untouched. He speculated that the mold might be killed by bacteria. In the course of his research to test this hypothesis, he also inquired whether the reverse might hold— could bacteria be killed by the mold? He inoculated guinea pigs with virulent cultures of typhoid and then gave some of them a broth from the penicillin mold. Those that received the mold recovered, he claimed, in contrast to those not treated, which died. Duchesne reported his research in a dissertation entitled "A Contribution to the Study of the Struggle for Existence of Micro-organisms: Antagonism between Molds and Microbes." But his work was totally ignored and was not rediscovered till the 1970s. Should Duchesne be given credit for the discovery of penicillin? What Fleming discovered was a particular strain (among several hundred in the genus *Penicillium*) that turned out to be unusually effective and well tolerated by most people. In cases where there is no continuity of earlier unrecognized work with later research, the scientific community generally accords little credit to the earlier work.

Suggestions for Further Reading

Cohen, I. B. 1985. *The Birth of a New Physics.*
Curie, E. 1937. *Madame Curie.*
Gibbs, F. W. 1965. *Joseph Priestley.*
Gillispie, C. C. 1970–1980. *Dictionary of Scientific Biography.*
Koestler, A. 1960. *The Watershed: A Biography of Johannes Kepler.*
Koyre, A. 1973. *The Astronomical Revolution.*
Macfarlane, G. 1984. *Alexander Fleming: The Man and the Myth.*
Schofield, R. E. 1966. *A Scientific Autobiography of Joseph Priestley, 1733–1804.*
Sootin, H. 1959. *Gregor Mendel: Father of the Science of Genetics.*
Westfall, R. S. 1980. *Never at Rest: A Biography of Isaac Newton.*
Wheeler, L. P. 1962. *Josiah Willard Gibbs.*
White, M., and J. Gribbin. 1994. *Einstein, A Life in Science.*

5 Scientific Reputation, Scientific Influence, and Public Fame

We will now examine some of the factors, apart from pure intellectual curiosity, that motivate scientists in their professional work. We will start in this chapter with two that can be described as internal to the scientific profession: the desire for scientific reputation, that is, for the respect of one's peers for the work one has done, and the ambition to exercise influence on the development of science and the way it is practiced. We also consider here the related, but distinguishable, ambition to achieve fame in the eyes of the general public. Other factors that might be described as external to the scientific profession are examined in subsequent chapters.

5.1 Scientific Reputation

Ambition to gain a scientific reputation is not in itself an unworthy motive—indeed, it is an important factor in the progress of science. Scientific knowledge must be publicly communicated. The criterion for establishing the reality of a reported phenomenon is its observability by other people when the required conditions are reproduced. For a theory to be recognized as scientifically valid, the scientific community must be brought (by rational discussion) to agree that the theory provides the best available explanation of the phenomenon in question. Thus, making a contribution to science requires having a claim recognized as valid by other scientists. Very occasionally, a scientist working in isolation may make a vital discovery. Mendel is one of the rare examples of this sort, but he acquired from others his understanding of scientific method as well as some idea of what questions in biology were worth asking. No one, however brilliantly endowed, could work all that out from scratch.

Our present conception of science and its methodology has been built up over the course of centuries by many minds. The great founders of modern physical science—Copernicus, Kepler, Galileo, and Newton— were themselves educated in the theories and methods of their predecessors. If someone works without benefit of discussion with competent peers and thinks of herself as having made important discoveries, she will have to communicate these findings (if only to a later generation) through writing. Perhaps such an isolated scientist could be confident enough of the validity of her results to feel personal satisfaction from the fact that she has discovered something; conceivably, she might also be so self-sufficient as not to care very much whether others recognized her work or not. Mendel appears to have satisfied this first condition but not the second, for he did publish papers, and he sought feedback in his correspondence with Naegeli. Newton, as we shall see, often shied away from public discussion, while at other times he seems to have cared excessively for his own reputation.

Isolation and self-sufficiency are of course extremely unlikely under present conditions of instant, around-the-world electronic communication. The normal context of present-day scientific research is a thorough preparation in the fundamentals of the subject, leading to familiarity with recent research in some specialized area, with rapid publication of any new results and almost immediate critical testing by relevant specialists all over the world. Those working at the frontier of a scientific field quickly get to know who is presently making the most important contributions.

One special feature attending scientific reputation, however, is characteristic of science as it has come to be practiced—namely, the great importance attached to *priority in discovery.* The first person to publish a result gains practically all the credit, even if several others have been working on the same problem and reach similar conclusions independently, perhaps publishing them only a short time later. It is a case of "winner takes all": he or she who first publishes a scientifically satisfactory account of X is usually known for the rest of history as *the* discoverer of X.

Historians of science may later try to set the record straight, pointing out cases of earlier discovery that did not gain credit in the scientific literature. In fact, it is not always easy to say who really made a discovery. Sometimes there is an anticipation of a later theory that is not developed much at the time, such as Aristarchus of Samos (ca. 260 B.C.) saying that the earth moves around the sun, thus foreshadowing Copernicus's heliocentric hypothesis. Tardiness in publication may allow the honor to pass to another (Joseph Henry is a case in point, as we shall see later in this chapter.) Swedish chemist Karl Scheele separated oxygen a couple years before Priestley and Lavoisier—he called it "fire-air"—but failed to publish first. A correct theory may be rejected at the time, and if it is not pursued, it generally receives little or no credit. A tragic case in point is that of J. J. Waterston, who worked out a fairly comprehensive kinetic theory of heat in an 1845 paper rejected by the two referees of the Royal Society as "mere nonsense." The potential value of the paper was noticed only as recently as 1892 by Lord John Rayleigh, long after Waterston had died and others had worked out the same ideas. Conversely, credit has always been given to William Herschel for the discovery of Uranus in 1781. But at least seventeen observations of the planet had been recorded before, though it was taken to be a star (Kuhn 1977, 171). And Herschel himself first thought it was a comet until the Russian astronomer Anders Lexell (1740–1784) suggested that it was the seventh planet. A more recent example is the anticipation of Fleming's discovery of penicillin, mentioned in 4.3.

Ever since science got going in the seventeenth century, scientists have been avidly competing for priority in discovery, sometimes breaking out into bitter disputes, as we shall see. In this century, the prestige attached to Nobel Prizes has probably exacerbated this tendency. The system of awarding these prizes (which are sometimes shared, but between at most three persons) requires the judges to make discriminations where it may be difficult to do so with justice. This tends to encourage an attitude that Einstein disparagingly compared to the competitiveness of athletics (see 3.2), with scientists vying to display their prowess not just by wrestling with the secrets of nature but, as it were, with each other. Just as there is only one winner in a race, even if the margin of victory is only a hundredth of a second, so it is with scientific priority as it has come to be interpreted and rewarded.

Why should scientific recognition follow a "winner takes all" rule? Many practical decisions have to be of an all-or-nothing type, for example, marriage, accepting a job, or voting. In a scientific department, there may be only enough money for one position or one research grant at a time. In such situations, choices may have to be made based on the narrowest of differences. But it is not obvious that in less immediately practical decisions, such as evaluating scientific contributions and reputations, we should expect there to be a single winner. Why place so much emphasis on being the *first* to do something? The way the credit for discovery gets awarded in science does not follow our usual notion of just deserts. In a study of how priority has been awarded in biology, David Hull noted that "the usefulness of scientific patron saints is not influenced very strongly by matters of intellectual justice" (1989, 124). The winning of prizes is not the primary incentive driving scientists (or artists, musicians, and novelists), but it may speed up their work, exacerbate its competitive element, and channel it in certain directions.

The awarding of priority strictly on the basis of publication seems to have developed as a way of getting the attention of the scientific community. This function was recognized by the pre-Darwinian biologist

Jean-Baptiste de Lamarck (1744–1829): "It is not enough to discover and prove a useful truth previously unknown . . . it is necessary also to be able to propagate it and get it recognized" (from *Zoological Philosophy*, 1809, quoted in Hull 1989, 124). Darwin expressed a similar conception in discussing his own lack of credit for an idea that German naturalist Ernst Haeckel (1834–1919) stated in 1868: "Ontogenesis . . . is a short and quick recapitulation of phylogenesis" (1876). Darwin sounded a bit disappointed that "no notice was taken" of this idea as mentioned in his book *The Origin of Species*, published in 1859, yet he conceded the point in scientific etiquette: "It is clear that I failed to impress my readers; and he who succeeds in doing so deserves, in my opinion, all the credit" (1899, 72). As we note later on, this marks the chief reason why Darwin, rather than Alfred Wallace, is acknowledged as the "father" of evolution by natural selection.

Most scientists do care deeply about their reputations among their scientific peers. A number of admirable qualities are needed to succeed in scientific work at the highest level—imagination, determination, sufficient confidence in one's own ideas for the effort to validate them to seem worthwhile, perseverance through all manner of difficulties (theoretical, experimental, financial, practical, and personal), and single-minded concentration on the job (with consequent neglect of many other matters). But these can be associated with or turn into the less admirable qualities of overconfidence, egotism, dogmatism, and a will to dominate and denigrate others. Care for reputation or fame can become an unreasonable obsession. Let us now examine a number of cases to illustrate the varieties of pursuit of scientific reputation. Where the story is unattractive, we tell it not to expose the failings of otherwise great individuals but to show by example just how subtle is the mix of a scientist's motives and how the balance can get upset.

Newton's Bitter Priority Disputes

Newton achieved a very high scientific reputation very quickly. But despite being recognized for his great contributions, he became involved in a series of nasty quarrels with other scientists, disputing priority even when his own position was secure. It remains true that his work was driven primarily by the sheer desire for knowledge and understanding of the inner workings of nature—which is the way we have portrayed him in 4.1. Nevertheless, Newton presents us with an interesting and somewhat disturbing case of overconcern with reputation.

As a student at Trinity College, Cambridge, Newton rapidly developed his precocious mathematical talent, impressing his tutors and making himself master of most of the new mathematics and science of the seventeenth century. He was soon elected a Fellow of Trinity College and became Lucasian Professor of Mathematics at the early age of twenty-six. From then on, he was assured the income and leisure he needed to do the research that was the abiding passion of his life. He pursued his studies with that single-minded determination that we have seen to be characteristic of the successful scientist. Newton readily achieved national and international recognition. In 1672 he became a fellow of the Royal Society, and he later attained the very rare distinction of election as a foreign member to the French Academie. He remained at the forefront of research in mathematics and physics for the rest of his life, with his eminence recognized by the leading scholars in Europe, such as Christian Huygens and Gottfried Leibniz.

Newton's life was a story of great success in scientific achievement and in recognition by the members of his profession—and toward the end of his life, even public fame, as we shall see in 5.3. Yet Newton also revealed a troubled inner psychology in his disputes with other scientists. It is difficult to uncover his personal feelings, especially at this distance of time, but we can get some clues from the strange attitudes he displayed in contentious situations. New ideas in science are not always received with instant acclaim, especially when they come from young and so far unknown figures. Today we are more used to the idea of scientific advances being frequently made, but in Newton's time even his majestic achievements faced a battle for acceptance. His reaction to scientific controversy seems to have veered from one extreme to another. In 1676, when he was pressed to reply to Linus, a Belgian scientist who had disputed some of his experimental results with light, apparently without having tried to reproduce them, Newton responded peevishly: "If I get free of Mr. Linus's business I will resolutely bid adieu to it eternally, excepting what I do for my private satisfaction, or leave to come out after me; for I see a man must either resolve to put out nothing new, or to become a slave to defend it" (Andrade 1954, 64–65).

This statement might express no more than passing exasperation with incompetent criticism. But in view of further evidence to be presented in the pages that follow, it appears indicative of an aversion to getting involved in any public discussion. It is as if Newton did not feel the need for others to check on his work and was unconcerned with making a professional reputation. If so, he was a very unusual scientist indeed. Yet in other, more serious matters of debate Newton went to the opposite extreme, getting provoked into disputations and long-lasting personal animosities that obsessed and blighted his last years.

The first of Newton's notorious priority disputes was with Robert Hooke (1635–1703), an experimentalist who made some wide-ranging contributions to early science. Hooke was appointed curator of the Royal Society, with the duty "to furnish the Society with three or four considerable experiments" at their weekly meetings. His treatise *Micrographia* of 1665, famous for the first description of cells, which Hooke observed in cork tissue, included results of many experiments on light as well as a hypothesis about its vibratory nature. Hooke had a fertile mind, but he jumped from one topic to another without developing any of them in depth. Apparently jealous of the celebrity of Newton's previously mentioned work on light, Hooke accused him of plagiarizing his own ideas. Newton repelled this charge in a crushing refutation but then refused to give Hooke any credit whatsoever for what he had done. The practice of acknowledging the contributions of other scientists was not then as standard as it is now, but Newton was unwilling to concede anything to Hooke, even when diplomacy might have suggested doing so. Reconciliation was attempted by mutual friends, but feeling ran high on both sides. Part of the blame for the dissension belongs with the secretary of the Royal Society, who disliked Hooke and promoted discord between the two at every opportunity. Newton became so disgusted by the whole affair that he resolved not to engage in any further correspondence on scientific matters or to communicate his results to the Royal Society. That is why his *Opticks* was not published until 1704, after Hooke's death. Here is an extraordinary case of a scientist deliberately cutting himself off from communication—quite the opposite of Mendel (see 4.2), an isolated scientist yearning for discussion of his work.

Newton was reluctant to make his work public and had to be persuaded into publishing his epoch-making *Principia.* He had laid the mathematical foundations for this work in earlier years but then put them aside to spend a decade in reclusive concentration on chemistry, alchemy, history, and theology. (He left behind some 650,000 words of writing on alchemy and over a million on biblical and theological topics.) His mathematical genius was sparked into renewed activity by a problem put to him by Edmund Halley (after whom the comet is named), who asked how Kepler's laws of planetary motion could be derived from the inverse square law of attraction. Newton applied his "method of fluxions" (the differential calculus that he invented) to the principles of dynamics to solve the problem.

When publication of *Principia* was at last imminent, Newton was so enraged by another claim to priority by Hooke that he threatened to withdraw the crucial third volume. He was ready to mutilate his own magnum opus simply because a lesser man had unjustly slighted him. Hooke does seem to have anticipated the idea of deducing the elliptic planetary orbits from the inverse square law, but he never properly developed it. At a meeting of the Royal Society in 1684, he claimed to have a deduction of the orbits "but would conceal it for some time so that others, trying and failing might know how to value it, when he should make it public" (Fauvel 1988, 45). Hooke's reputed boasting whenever Newton discovered something that he had already discovered it must have been very irritating to the sensitive Newton. It was the actual derivation of Kepler's laws from the law of gravitation that was Newton's great achievement. All of Halley's diplomacy was needed to persuade the enraged Newton to let the world see his work after all.

Another of Newton's quarrels became one of the most famous of all scientific priority disputes, which was between Newton and Leibniz over their respective contributions to the invention of calculus, an indispensable tool in physics that today finds application in almost every scientific field. The dispute had its roots in Newton's early work, but it became a public, even international, cause célèbre in the last period of his life, rousing him to even greater fury than previous quarrels. Examination of the papers left by the two principals has shown that Newton did have priority of discovery, arriving at his "method of fluxions" by 1666, about ten years earlier than Leibniz's independent invention of differential calculus.

Both men deserve immense credit. But they took very different attitudes toward their discoveries. Newton was secretive about his mathematical work, not publishing anything important until 1704, when he was sixty-one. Leibniz, who needed to earn his living by his writing, published quickly and later tried to get the undivided credit, despite the fact that he had had the benefit of some correspondence with Newton. Leibniz even complained to the Royal Society about not being given proper recognition. In 1712, a committee of the society investigated the dispute and reported in Newton's favor. It was only discovered many years later that Newton himself, as president of the society, had written part of the anonymously published report of the supposedly impartial investigation. Egged on by their supporters in each nation, neither of these great men was prepared to share the fame.

Henry: Not Publishing in Time

Joseph Henry (1797–1878) was the greatest American experimental scientist of the nineteenth century, but during his lifetime he was better known as an able administrator of the Smithsonian Institution. He was in fact the first to hit on the principle of electromagnetic induction, but the great English experimentalist Michael Faraday (1791–1867) made the same discovery a little later but published first and thus has been given the credit.

Henry's background was rather similar to that of his more famous contemporary. He came from a poor family, received little schooling, and went to work at an early age. Like Faraday, his potential interest in science needed only a small spark to ignite it. At age sixteen, so the story goes, while visiting a relative's farm, he chased after a pet rabbit under a church and found his way inside through some loose floorboards. Inside he found himself in the village library. Returning there several times got him interested in books. Later he came across a collection of *Lectures on Experimental Philosophy, Astronomy, and Chemistry,* which awakened his scientific curiosity and inspired him to return to school. He began experimenting with electromagnets in his spare time, eventually making one that could lift more than a ton of iron; he used his wife's silk petticoats for insulation material. His invention of an electric relay was the principle behind the telegraph developed by Samuel Morse. In 1831, he built an electric motor powered by batteries. This motor, which he considered a "philosophic toy," became the basis for the great variety of electric devices in use today (Coulson, 1950, 70). Henry never tried to patent his inventions, as he believed that one should not profit from scientific work. In 1842, he demonstrated in effect the transmission of radio waves a half century before German physicist Heinrich Hertz (1857–1894), his report on this discovery showing that he understood it as a wave phenomenon.

It was in 1830, however, that Henry made his greatest discovery. Previous investigators of electric phenomena were misled by the fact that a steady current induces a steady magnetic field. Henry noticed that to induce a current, a changing magnetic field is required. He formulated the principle of induction that describes how an electric current in one coil induces a current in another by means of its magnetic field. But his heavy teaching load in mathematics and science at the Albany Academy and later at Princeton didn't give him enough time to publish right away. In the meantime, he read in a British journal that Faraday had made the key experiment—after Henry, but beating him to publication. Though quite disappointed, he never contested Faraday's priority. As he put it: "I ought to have published but I had so little time. . . . I wanted to get out my results in good form" (Coulson, 1950, 87). He was persuaded to write up his results anyway, including the principle of self-induction, for which he did have priority. One reason his work was not much recognized was that U.S. science did not have a very high reputation at the time. He did finally achieve some posthumous fame when the unit of inductance was labeled the "henry."

Darwin and Wallace: A Question of Priority

One of the great scientific revolutions takes its name from Charles Darwin (1809–1882). He was not the first to propose evolution, but his voluminous writing established the theory of evolution by natural selection as essential to the foundation of biological science thereafter. The beginning of the Darwinian revolution is dated from the appearance of *The Origin of Species by Means of Natural Selection* in 1859. Darwin had been developing his ideas ever since his expedition around the world on the British naval surveying ship the *Beagle* (see 7.1). On his return to England in 1836, he began keeping notebooks, working, as he remarked, "on true Baconian principles" by collecting facts. Sparked by reading a passage by English economist Thomas Malthus (1766–1834) on population pressure leading to a struggle for existence, he hit on the main idea of his theory. He began work on a big "Species Book" that would demonstrate the workings of natural selection in giving rise to the diversity of living nature with all its wonderful adaptations.

Darwin held off publishing his ideas on evolution for many years, partly from fear of the public response—the matter was (and still is) intensely sensitive in theology and was thought to be subversive in morals and politics. He shared with Newton a distaste for public controversy. However, his friend, pioneering geologist Charles Lyell (1797–1875), had warned him that his ideas might be anticipated and that he had better publish. Evolutionary notions were very much in the air in nineteenth-century England. The general idea of natural selection had already been proposed by several writers, but without being developed as a full-blown theory of evolution. Darwin confessed, "I rather hate the idea of writing for priority, yet I certainly should be vexed if any one were to publish my doctrines before me" (quoted in Irvine 1955, 98). He was moved to begin a sketch of his theory but found it difficult to condense his whole line of argument. It took a letter in 1857 from another biologist, Alfred Russel Wallace (1823–1913), to stir him into action. Wallace had independently arrived at natural selection as a mechanism for evolution. Darwin was stunned, and he wrote to Lyell:

> Your words have come true with a vengeance. . . . I never saw a more striking coincidence; if Wallace had my MS. sketch written out in 1842, he could not have made a better short abstract! . . . Please return me the MS., which he does not say he wishes me to publish, but I shall, of course, at once write and offer to send to any journal. So all my originality, whatever it may amount to, will be smashed, though my book, if it will ever have any value, will not be deteriorated; as all the labour consists in the application of the theory. (Darwin 1899, vol. 1, 473)

The idea of natural selection had occurred to Wallace while he was on an expedition gathering specimens in Indonesia. He had read Darwin's *Beagle Journal* and had been pondering the question of transmutation of species. During a bout of fever, "in a flash of insight" he hit upon the idea of natural selection, that the fittest would survive. Like Darwin, Wallace had been stimulated by the population theories of Malthus, whose work he recalled having read twelve years earlier. He dashed out his ideas in the short letter that he sent to Darwin, with whom he had previously corresponded about evolution. Darwin must have been tempted to ignore Wallace or put him off while he rushed into print, but he was not up to doing so. The conscientious Darwin at one point even considered withdrawing in favor of Wallace, exclaiming that "I would far rather burn my whole book than that he or any other man should think that I had behaved in a paltry spirit" (Desmond and Moore 1991, 469). But he was not really ready to chuck twenty years of labor. He appealed to his friends Lyell and botanist Joseph Hooker (1817–1911), who were familiar with his work, to come up with an honorable solution. They arranged for Darwin to present a summary of his work along with Wallace's paper before the Linnaean Society in 1858. This solution to the problem of credit for independent discovery has provided a model for later cases—which come up remarkably often.

Why then do we now speak of Darwin's, not Wallace and Darwin's, theory of evolution? There are some subtle differences in their conceptions of natural selection, but this is not the reason for Darwin having gained preeminence. Darwin had thought of the idea before Wallace, but this alone, without publication, does not count for much in science. Surprisingly, the papers of 1858 did not arouse much interest. Only after the publication in the following year of *The Origin of Species,* a large work with a mass of detailed evidence supporting the theory of evolution by natural selection, did the public and the scientific community begin to pay attention. Wallace did some further work on evolution, but his retreat in later years from the idea that natural selection could account for the human mind earned Darwin's scorn, and eventually Wallace lapsed into spiritualism (like so many in the Victorian age). He seemed quite content to let Darwin get the scientific credit, and in 1889, he even entitled his own popular account of evolution *Darwinism.* About his own role he wrote Darwin, "All the merit I claim is the having been the means of inducing *you* to write and publish at once" (Marchant 1916, 131).

Watson: The Race to Crack the Code of Life

James Watson, Francis Crick, and Maurice Wilkins shared the Nobel Prize in physiology in 1962 for their "discoveries concerning the molecular structure of nuclear acids and its significance for information transfer in living material." The discovery of the helical structure of DNA ranks as one of the greatest scientific breakthroughs of the twentieth century, for it has resulted in our understanding the very code of life and was the beginning of the field of genetic engineering, which now promises—or threatens—to radically affect our lives. Their research showed that genes, the hereditary factors postulated almost a century earlier by Mendel, functioned by a kind of chemical lock-and-key mechanism. Though everyone agrees on the momentous nature of the result, the process of the discovery itself has been clouded by controversy.

Watson brought much criticism on himself by writing a remarkably frank account of the process, which reveals the highly competitive nature of modern science. *The Double Helix* (1968; reprint 1980) is one of the few descriptions from within of what it is like to do scientific research at the cutting edge of knowledge. Watson conveyed the intellectual fascination with a crucial problem, an excitement bordering on obsession—though not excluding enjoyment of tennis, wine, and the company of the pretty girls in Cambridge (no austere monastic scientist this!). He described the rivalries between different groups working on the problem; the frustration caused by adverse experimental results, unreliable texts, and recalcitrant lab equipment; worries about funding and about his standing with the lab boss; and difficult colleagues. Not surprisingly, the book became a best-seller.

Watson was a whiz kid from Chicago who had entered college at fifteen and earned a Ph.D. in genetics at Indiana University by the age of twenty. Postdoctoral fellowships enabled him to study at Copenhagen and then at Cambridge, England, from 1951 to 1953. To some of the more staid British, he seemed like a brash, intellectually arrogant cowboy. Watson breezed in at a time when it was becoming clear that DNA (deoxyribonucleic acid) must somehow be involved in the transmission of genetic information from one generation to the next in every form of life. Before the 1950s, most biologists had not expected that the mechanisms by which genes operate would be discovered soon. A few, however, such as the charismatic Linus Pauling in California, were racing to make the breakthrough, realizing the possibility that unraveling the structure of DNA might reveal how information is transmitted when cells replicate. In King's College, London, Maurice Wilkins and Rosalind Franklin had been doing careful experimental work on the X-ray diffraction patterns from DNA molecules.

There was at the time a certain informal ethic or etiquette in scientific research in Britain that those who had started work on a topic should be allowed a reasonable chance to make progress, publish results, and get credit before others had a go at the problem. Watson, however, was not one to be discouraged by the disapproval of Sir Lawrence Bragg, the distinguished director of Cambridge's Cavendish Laboratory, who held old-fashioned, genteel views about his staff muscling in on other peoples' research topics. After all, nobody could prevent him from thinking! Watson benefited from the eager collaboration of the slightly older but as-yet-unestablished Francis Crick, a former physicist whose flair for biochemical theory was crucial in this and subsequent discoveries. But Watson himself obviously had both a nose for the deeply significant questions of genetics and a determination to pursue them despite administrative, financial, and personal obstacles. He displayed some shrewd deviousness (sometimes necessary for success in science, as elsewhere!) in bending the rules governing his postdoctoral fellowships so that he could concentrate on the DNA problem rather than the somewhat different research for which he was supposedly being paid. (He had applied for a National Science Foundation grant but was rejected, which perhaps says something about the unpredictable nature of important research.)

Watson and Crick have been accused of unethically using the results of X-ray crystallography, painstakingly achieved by Rosalind Franklin, to construct their model of DNA and thereby win the race for publication. Watson indeed wrote in his book of sitting in the back of Franklin's seminars in London to pick up the latest experimental evidence about DNA, without letting on to her that he and Crick were also working on the problem. But at one stage he failed, in his youthful overconfidence, to take adequate notes of certain details that turned out to be important. In 1951, Watson attended a lecture in which Franklin inferred from her X-ray photographs of DNA that the position of its phosphate groups were outside the helix. Watson and Crick had for two years tried unsuccessfully to model DNA with these groups *inside* the helix. An announcement by the Cambridge pair that they had got a structure for DNA brought Wilkins and Franklin on the next train from London to inspect the model that Watson and Crick had painstakingly constructed. But the London pair were able to point out mistakes straight away. Bragg then thought his own opinion confirmed—the two young mavericks had been wasting their time, as well as violating scientific etiquette, and he banned them from further work on the topic. (Crick was firmly instructed to get back to his unfinished Ph.D.) But they could not be prevented from talking to each other about the hot topic of DNA, and after another fifteen months they came up with the correct model: a double helix, with the four kinds of nucleotide building blocks. The nucleotides differ in their nitrogenous bases: adenine (A), guanine (G), thymine (T), and cytosine (C). The four bases bond in pairs, A to T, G to C, like rungs in a ladder. The next time Wilkins and Franklin were invited to Cambridge to inspect a model, it had been very thoroughly checked beforehand, and the experimentalists accepted its correctness. Shortly thereafter in 1953, the historic brief paper was published in *Nature*.

A Nobel Prize was awarded in 1962 to Watson, Crick, and Wilkins. Franklin had meanwhile died of cancer and so was not eligible according to the Nobel rules, which do not permit posthumous awards. According to Crick's subsequent estimate, Franklin and Wilkins would have shared a second prize if Franklin had lived. But the question of the relative contributions of the two pairs of scientists remains controversial, not least because it raises feminist issues. Anne Sayre set out to put the record straight in her study entitled *Rosalind Franklin and DNA* (1975), described on the cover as "a vivid view of what it is like to be a gifted woman in an especially male profession." Sayre concluded bluntly that "Rosalind has been robbed" (190), quoting from Robert Frost's poem "Kitty Hawk": "Of all crimes the worst / Is to steal the glory." She also compared the acknowledgment of a scientist's work as analogous to claiming parenthood for one's children. Sayre suggested that a joint paper with Franklin on the structure of DNA fairly representing the various contributions would have been more appropriate than having Watson and Crick hog the glory, as she put it. She also argued that the case set a bad precedent; if competition in research is uncivilized and unrestrained, "the rush to publish before someone else . . . is a rush in which poor work done in haste is encouraged" (195).

Watson did present an unflattering, even cruel, portrait of Rosalind Franklin in *The Double Helix*. "Rosy," as he called her, was unkempt, bitchy, and impossible to deal with—at least, this was how she appeared to him at the time. Other men recalled Franklin as being "difficult." Wilkins once remarked that she was "very fierce, you know" (quoted in Bernstein 1978, 147.) Part of the problem seems to have been that she was the only woman in a male-dominated department, with a particularly strong prejudice in Britain against women going into science at all. Women were even denied access to the Staff Common Room in King's College, London, during that era. Franklin had an independent income and, hence, the opportunity to do many other things with her life, but she deliberately chose, against the advice of her family, to dedicate herself to scientific research.

Watson and Crick stole Franklin's crucial data in the sense that they used it without her consent. The question of whether Watson and Crick violated scientific etiquette—by failing to properly cite her contributions—is more complex. As with most scientific research, the discovery of the structure of DNA depended on many forerunners, including the work of Erwin Chargaff on nucleic acids. Linus Pauling's work in 1948 on the helical structure of the protein keratin provided a basis for hypothesizing a similar structure for DNA. Wilkins had already tried a single-helix model before Franklin came to work at King's College. Sayre claimed that it was Franklin who initiated the attempts to demonstrate a helical structure using X-ray photographs. She and her assistant Raymond Gosling did continue work, started by Wilkins, that Watson and Crick later used to construct their model. But there had already been a quarrel between Wilkins and Franklin that resulted in a lack of communication between them, even though they worked in the same department. Franklin and Gosling failed to publish any of their work on DNA structure until 1953, treating it "as private and personal data" (Bernstein 1978, 154).

Physicist and writer Jeremy Bernstein (1978) suggested that the one most hindered by this fact was Pauling, who had begun looking into the structure of DNA in 1952 but had only older, inferior X-ray data available to him. Franklin does appear to have abandoned her initial hypothesis of a helical structure for DNA in 1952, not coming back to the correct conception until January 1953, after Watson and Crick had already constructed their double-helix model. They did actually offer to collaborate with the King's College group at one point, but Franklin and Gosling were not interested. As for proper citation, Watson and Crick didn't cite data at all in the paper in *Nature* announcing their discovery—the data were later published by Wilkins and Franklin. Watson and Crick did include in their paper a weak acknowledgment of their debt, saying they had been "stimulated by a knowledge of the general nature of the unpublished experimental results and ideas of Dr. M. H. F. Wilkins, Dr. R. E. Franklin, and their co-workers at King's College" (Watson 1980, 241). The greatest unfairness, Bernstein suggested, was that the London team was not informed of the use being made of their data. He viewed Watson and Crick's treatment of Rosalind Franklin as simply common practice in science (note the assumption of male gender): "To a working scientist, what has not been published must be assumed not to exist. Part of the job of a serious scientist is to publish what he believes in. If he fails to do this and someone later rediscovers his work, one may admire the first man for his ingenuity, but, in most cases, he is out of luck" (1978, 161). Even if this describes the usual "ethics" of scientific priority, it may not always represent justice.

Bell: Sex Discrimination in Science

The case of Rosalind Franklin not being given due credit for her work has been used as a paradigmatic example of sexist discrimination in male-dominated science. Watson exhibited a great deal of stereotypical chauvinism in his description of Franklin and in downplaying the importance of her data. But the question of how her contribution to the discovery of the structure of DNA should have been credited is, as we have seen, rather complex and involves more than gender-based discrimination. A rather more straightforward case of such discrimination can be found in the treatment of Jocelyn Bell in the discovery of the type of stars known as pulsars.

Jocelyn Bell (later Dr. Burnell) became interested in astronomy as a child. After studying physics, she decided to work in radio astronomy for her doctorate because, in contrast to optical astronomy, you didn't have to stay up late at night to make observations. She joined an astronomy group led by Antony Hewish at Cambridge, England, in 1965. Hewish had designed a telescope to study the effect that the solar wind (particle radiation from the sun) had on a "twinkling" of radio-wave emissions of stars. The radio telescope swept the sky once every four days, producing hundreds of feet of charts. Bell was set to work distinguishing stellar sources of radio signals from earthly sources. In October 1967, she noted a peculiar signal on a half-inch portion of a 400-foot chart. She recalled having seen what she described as "a bit of scruff" on the chart in the same part of the sky some time earlier. Bell discussed the observations with Hewish, and they decided to study them further.

For a time no more such bits of scruff appeared, and Hewish was about to dismiss them as background "noise" or recordings from a flare star. Then Bell found pulses coming in very regularly at 1.5 seconds apart—which suggested a human source, except that the pulses kept "star time" (a sidereal day of twenty-three hours and fifty-six minutes, as opposed to twenty-four hours). That December, Bell was working one evening (despite her dislike of night work) on analyzing a chart. She saw more "bits of scruff" in a part of the sky that was to be passed through again at 1 A.M. The right stuff came through for five minutes: a train of pulses now 1.25 seconds apart. She left a note on this for Hewish and went on a vacation. With repeated observations of the unusual signals, Bell and Hewish "facetiously entertained" the idea that the source might be "Little Green Men" from some extraterrestrial civilization that was trying to communicate with us (Judson 1980, 96). The first pulsar radio source was actually referred to as LGM-1. From this data, Hewish finally concluded that they must be dealing with a rapidly rotating star. He labeled it a "pulsating star," later shortened to *pulsar.*

Many observers would have passed off the scruff as noise, since radio telescopes are very sensitive to disturbances from a number of sources, such as auto ignitions. (Had Bell not been analyzing the data by hand—

this is commonly done by computer today—she would probably not have noticed the pulsar emission at all.) Bell told the story of a radio astronomer who, several years before the recognition of pulsar signals, had actually picked one up. He saw the pen jiggle but dismissed it as simply his equipment misbehaving. He kicked the table, and the pen stopped jiggling. Bell, however, investigated the strange signals further, checking whether the apparent pattern of radiation correlated with any known sources. She could not find any correlation between the scruff and earthly disturbances, and the recordings did not at all fit any known type of star radiation.

Bell's data eventually led to the discovery of pulsars, which are tiny neutron stars some ten miles in diameter but extremely dense—a volume of pulsar material the size of a pinhead may have a mass greater than a million tons. Pulsars send out a beam of radiation that appears to us something like the rotating beam of a lighthouse. The existence of neutron stars had been postulated on theoretical grounds in the 1930s as the result of a star undergoing gravitational collapse under certain conditions.

An article describing the pulsar signals was published in *Nature*, with Bell's name second on the list of five authors. Hewish, along with Martin Ryle, who had designed the first radio telescope, received the 1974 Nobel Prize in physics for the discovery. A number of astronomers protested that Bell should have been included in the prize, since it was her persistence in following up the anomalous radiation data, albeit in a project directed toward rather different problems, that led to the discovery. Astronomer Fred Hoyle said the award was a "scandal," noting that Bell's "achievement came from a willingness to contemplate as a serious possibility a phenomenon that all past experience suggested was impossible" (Broad and Wade 1982, 48). Hewish responded patronizingly that "Jocelyn was a jolly good girl, but she was just doing her job. She noticed this source was doing this thing. If she hadn't noticed it, it would have been negligent."

How much the lack of recognition in this case was due to gender and how much to the common practice of scientific superiors taking credit for their assistants' contributions may be difficult to say. It is also relevant that Bell's contribution was primarily at the observational level; theoretical innovation tends to carry greater prestige in science than observational skill. But this example supports the conclusion that "the supposed meritocracy of science rests on a power structure, and those who hold the reins of power are also influential in controlling the allocation of rewards and credit" (149).

5.2 Professional Power and Influence

The power we are concerned with here is that exercised within the scientific professions, usually as a result of special expertise. Many university scientists also have influential administrative positions as leaders of research teams, heads of departments, chairs of research councils, and so on. Some may even get to the point where they cease to be active scientists, enjoying their power more than any research. There are also those who undertake administrative responsibilities only from a reluctant sense of duty—as well as those who do their very best to avoid them! We will examine a variety of examples.

Newton's Unscrupulous Use of Power

Let us look at Isaac Newton again, this time as an example of scientific reputation leading to power within the profession. To some extent, professional reputation of itself gives influence, since the higher the respect for someone's scientific achievements, the more his views, conjectures, and opinions will be taken seriously. But formal power, in the sense of holding institutional positions, does not necessarily or immediately follow. In Newton's case, his secretive and prickly nature probably discouraged his fellow scientists from entrusting him with office. At any rate, he did not become president of the Royal Society until 1703, after his opponent Robert Hooke had died and long after his scientific reputation was established. Newton then held the post until his death, ruling the society with great solemnity and dignity (and an iron hand).

Newton's use of his exalted position in the scientific world is shown in another of the great quarrels of his life. This one involved John Flamsteed, first holder of the newly established position of astronomer royal. Having overcome many difficulties in setting up the Greenwich Observatory from scratch, Flamsteed's systematic observational work had produced a great catalog of the stars in the sky. Disagreement with Newton first arose over a small point: whether two celestial appearances observed in 1680 were of different comets or of the same one twice. Newton eventually had to concur with Flamsteed's view on this topic and mentioned the matter in his *Principia,* but without, Flamsteed felt, giving due acknowledgment to his work. Personal differences between the two then developed into a long battle for control over the publication of Flamsteed's observational data, which were crucial to the verification of Newtonian theory. The details reflect little credit on the adversaries. Suffice it to say that matters came to a stage where Newton used his influence at the royal court to get an order requiring Flamsteed to submit his material for editing and publication by a committee of the Royal Society, dominated by Newton and his supporters. Flamsteed had the last laugh, however, because after Queen Anne's death, he was able to reverse the situation, seizing and burning the undistributed copies of the edition that he felt so misrepresented his life's work and arranging for republication in the form he wanted (albeit at his own expense). Not many people were able to get the better of the imperious Newton, who by that time had added political influence to his almost godlike scientific status.

No doubt the story has its less dramatic counterparts in many scientific departments and, indeed, in most human institutions. It is a theme so familiar as to be banal: Someone attains a position of influence, quite deservedly on the basis of her performance, and then abuses her power to a lesser or greater extent. It is no surprise that examples should turn up in the practice of science, and it is only the more melodramatic cases that get public notice. The following is a more recent example.

Burt's Megalomania

Scientific fraud is a rare phenomenon—or so it has been assumed until recently (see Broad and Wade 1982). The way that scientific theories are tested tends to exclude it (see 1.1 and 3.1). Valid experimental results should be reproducible, so false claims can usually be detected by repeating the experiment. There is thus little prospect of getting away for very long with assertions one knows or suspects to be incorrect. Yet fraudulent claims have occasionally been made and accepted in science, and it is revealing to examine how this can happen. English psychologist Sir Cyril Burt (1883–1971) is probably now fated to be more famous for his fudging of data about the inheritability of intelligence than for his positive contributions to psychology, which were considerable. (During his lifetime, psychology became established as an empirical science, and Burt's work did much to bring this about, at least in Britain.) But just as the study of abnormal or diseased organs can show something about the normal functioning of the body, so examination of cases of scientific fraud may reveal something of the motivations behind scientific work generally.

Probably the strongest influence on Burt's approach to psychology was the work of Francis Galton (Darwin's cousin) on individual differences and heredity. In his early years, Burt devised practical tests of reasoning ability and applied them to Liverpool schoolchildren. When he was appointed to a post with the London County Council (L.C.C.) in 1913, he became Britain's first professional applied psychologist, and he was occupied for the next twenty years with educational psychology. He had to organize the testing of many thousands of children for immediate practical purposes. He could not do it all himself and had to rely on assistants, who were often untrained. Yet this was the source of much of the data on which his more important and controversial theoretical claims later had to rest.

Burt's conviction always was that educational achievement depended largely, if not entirely, on intelligence, which he tendentiously defined as "innate, general, cognitive ability" (1955, 158). This is controversial in several ways. Behaviorist psychologists contended that concepts of mental states such as *abilities* were useless for explaining behavior, and the meaningfulness of IQ (intelligence quotient) tests as measures of a supposed "general cognitive ability" continues to be seriously questioned. But his claim that intelligence is *innate*, that is, genetically determined, was especially controversial because Burt was influential in instituting the English educational practice of selecting pupils for different types of secondary school on the basis of an examination at age eleven. Burt was a firm believer in the innateness of most individual differences and in the wisdom of selective schooling.

His appointment in 1932 to a Chair of Psychology at University College, London, took him from practical to academic work and forced him to face up to methodological and theoretical questions. Here he made himself a master of factorial analysis, a method for handling a complex set of observations in terms of a limited set of underlying factors. Having taught himself the necessary mathematical techniques, building on the work of his predecessor Charles Spearman, he applied them to the topic of individual psychological differences. He thought that the resulting book, *Factors of the Mind* (1940), would be his most lasting contribution.

This is the point where evidence of an obsession about his scientific reputation and influence began to appear. In the second edition of that book and in later works, Burt withdrew his acknowledgments of Spearman's fundamental contributions and started trying to represent himself as the originator of the factorial method in psychology. His claims became increasingly exaggerated, and he even resorted to the device of writing fake letters from nonexistent people for publication in the journal he edited! The falsity of his claim to priority should have been obvious to those who knew the subject. So why did he attempt the deception? It looks as if his egotism got out of control, turning from the self-confidence that any original scientist needs into delusions of a grandeur beyond that which he had earned.

Despite these failings, Burt's professional reputation lasted his lifetime. His claims about the innateness of intelligence involved him in public controversy because of the support they gave to the politically sensitive policy of selective education. But he was an agile debater, well able to defend himself in argument. (It is quite possible for scientists to engage in political disputes without damaging their professional reputations.) Soon after Burt's death, however, grave doubts were raised about the soundness of his observational data—and about his integrity in presenting them. These are serious charges for any scientist to face. Professional incompetence is bad enough, but outright fraud is unforgivable in the ethics of science.

A crucial kind of evidence for or against claims about the heritability of intelligence comes from instances of that rather rare phenomenon: identical twins reared in different family and social backgrounds. Burt, having been so long in the job of testing the intelligence of huge numbers of London schoolchildren, had had a special opportunity for collecting such data. At various times—in 1943, 1955, and in the 1960s—he published papers, sometimes mentioning as assistants a Miss Conway and a Miss Howard, in which he presented data that had ostensibly been accumulated on separated identical twins in his years with the L.C.C. (or later). But in 1972, Leon Kamin showed just how untrustworthy these so-called data were. No clear details were given of how they were collected; contradictions existed between the different published versions of them; careless errors could readily be detected; and—most damaging of all—extremely improbable tables of data were presented, with correlation coefficients coming out the same to three decimal points for samples of quite different sizes. At best, this shows such carelessness as to amount to incompetence; at worst, it looks like deliberate misrepresentation of the alleged data.

Hearnshaw's careful review of the evidence in his biography *Cyril Burt, Psychologist* (1979) supports the graver charge. It seems that Burt did no new surveys of twins after leaving the L.C.C., contrary to what he sometimes claimed, and that if Miss Conway and Miss Howard ever existed, they did no work for him after 1940. So either he was extremely forgetful, or he lied about the provenance of his data. What then *was* the source of the figures he published? In fact, many of Burt's records were destroyed in the wartime blitz on London, and the most probable conclusion seems to be that in his later years he was trying to bolster his controversial arguments by "reconstructing" lost or fragmentary or half-remembered data. Perhaps he continued to believe that he had *formerly* gotten empirical evidence for the heritability of intelligence, but he appears to have lost all sense of standards of honesty in arguing a scientific case. If his data had become incomplete or were based only on memory, he should have frankly acknowledged this.

Why then should such a highly intelligent man, who had proved to be a very competent psychologist and had occupied high positions of public trust, have stooped to dishonesty? His belief in the genetic determination of intelligence seems to have become so dogmatic that he adjusted evidence to fit his theory. Self-confidence turned into egotism, and scientific ambition degenerated into a craving for reputation and educational influence beyond what he had any right to expect. Normal scientific motivation can become pathological.

In Burt's case, several things went wrong for him in midlife, which may help explain the change. His marriage broke down; many of his important papers were destroyed in the war; and on top of all that, he developed severe symptoms of Meniere's disease, in which disturbance of the semicircular canals causes attacks of vertigo and impairment of hearing. There were professional disappointments, too: seeing his department taken over after his retirement by someone with quite different views; losing the editorship of his statistical journal; and the rejection by the Labour government (after 1964) of the policy of selective schooling with which he had been so closely identified. He must have felt that the world was rejecting all that he had stood for. Given all these circumstances, it is understandable that he became somewhat paranoiac.

Burt's is a rather clear case of fraud, but it is an unusual one even within the known cases in science. He appears to have become psychologically disturbed in a way that destroyed both his scientific integrity and his judgment. Ironically, nothing has cast so much doubt on the significance of genetics for intelligence, and even on the ability of psychologists to define it and test for it, as Burt's betrayal of scientific honesty in his IQ studies. Much more common in science are cases of slight "cleaning up" of data, where absolute judgments don't apply so clearly. Data-fudging has been attributed to some of the greatest scientists (though not without controversy—see Hull 1988, 313–319). Galileo is supposed to have made up some of his data on balls rolled down inclined planes, Newton is suspected of having fudged some of his calculations on gravitational forces, and some of Mendel's data, from which he calculated ratios of dominant to recessive characters in peas, is thought to be too perfect to be statistically plausible. But these cases are rather different from Burt's downright fraud. Standards of reporting experiments were less rigidly codified in earlier times, there was no gross deception in those examples, and the claimed results have been borne out by repeated testing. Broad and Wade apply a rather simplistic black-white notion of objectivity and honesty in science when they claim that "for the moralist, no distinction can be made between an Isaac Newton who lied for truth and was right, and a Cyril Burt who lied for truth and was wrong" (1982, 213). Nevertheless, it is important to recognize that scientists are constantly tempted to interpret their observations in terms of their favored theory, although the critical, skeptical attitude of other scientists can—but doesn't always—provide a safeguard against this understandable tendency.

We have seen how Burt's overconcern for reputation and influence seems to have led him into fraudulent claims. But before those were unmasked, there was other evidence that he had a strange cast of mind. An entry in his diary as an Oxford student reads: "My purpose in life concerns primarily myself. It is to produce one perfect being for the Universe" (Hearnshaw 1979, 268–269). Perhaps not too much importance should be attached to such youthful jottings, but in the context of the other evidence, this one seems significant. In later years, Burt harbored an intense dislike for his colleague Dr. Philpott, to whom he tended to delegate much of the routine business in the Psychology Department of University College, London. Personal antipathies are typical of human nature—even among supposedly objective scientists—but this was an extreme case. In one episode, the esteemed professor composed such a barrage of lengthy letters pouring scorn on the work of the luckless Philpott as to put Burt's own mental balance into question. After that, they were never on speaking terms again, and Burt vindictively blocked all chance of promotion for his unfortunate colleague.

Other devious and deceptive behavior appeared. Burt was wont to change the minutes of business meetings and to alter the proofs of papers without the author's consent so that they would fit his own point of view. After his retirement from his London chair, he tried to interfere with the work of his successor (of whom he did not approve), and the college authorities eventually had to take the extraordinary step of banning him from the campus! No doubt every institution has its horror stories to tell, but what we see in Burt's case is the common desire for professional power (even over relatively small academic and administrative details) magnified to a grotesque degree. It is sobering to note how difficult it can be to resist the authority of a scientist once he attains a high professional reputation along with a prestigious position.

The Baltimore Affair

The Nobel laureate David Baltimore has become known to the public not for his important work in immunology but for a scandal concerning a paper by him and five co-authors. A postdoctoral fellow in Baltimore's lab at the Massachusetts Institute of Technology (MIT), Margot O'Toole, working under the supervision of one of the co-authors, Thereza Imanishi-Kari, accused her supervisor of not having actually performed some of the experiments on which the paper was based. The article cited experiments in which antibodies expressed by one mouse could be made to mimic those of another mouse. O'Toole accidently came across Imanishi-Kari's laboratory notes and concluded that these records showed clearly that the experiment had not yielded the published results. She first questioned Imanishi-Kari but got no satisfactory explanation. She then told Baltimore her suspicions, but he dismissed them. She finally appealed to MIT officials, but the only action she got was getting fired. The most disturbing thing about the case was the extraordinary efforts of Baltimore and others in scientific authority to conceal the truth.

Baltimore himself has not been implicated in the original falsification of data, but it became clear that he was arrogantly unwilling to acknowledge clear evidence of malpractice in his own lab until forced to do so by a congressional investigation. University committees at Tufts and MIT rejected O'Toole's allegations, as did a National Institute of Health (NIH) panel (which, curiously, included a former student of Baltimore as well as a scientist who had co-authored a text with him). The fact that many in the scientific community naively rushed to the support of Baltimore, without any proper investigation of the case, does not contribute to public confidence in the objectivity of scientists in such cases. Professional loyalty seemed to dominate obligation to society. It took two NIH committees and three congressional hearings over three years to settle the case. Representative John Dingell, accused of holding a witchhunt, called in the Secret Service, which used sophisticated methods to show a clear pattern of falsified data. Only then did Baltimore ask that the original paper be retracted. But as often seems to happen, the whistle-blower was left without her job, while the accused Imanishi-Kari won a prestigious appointment.

5.3 Public Fame

Achievement in science sometimes leads to wider public fame, but such fame is not always proportionate to the individual's strictly scientific distinction. It is not very realistic to become a scientist with the aim of gaining worldwide celebrity, but a number of people have attained it, whether they wished it nor not. If celebrity comes, there are various attitudes scientists can take to it. Some may regard it with lofty indifference, thinking of it as a handicap to their work because it brings external pressures such as the attentions of the media—though they may find it an advantage in securing funding for further work. Some may rather enjoy a bit of public adulation. Thus, scientists can shun publicity, exploit it, or even revel in it. Let us look at some contrasting case studies. The reader might also consider this angle while looking at the cases of M. Curie in 4.2, Pasteur and Salk in 6.1, Darwin in 7.1, and Oppenheimer in 9.1.

Newton's Entry into Public Life

Newton is an example of a scientist who was able and willing to turn his great scientific reputation into fame and fortune of a more worldly kind. By the time his *Principia* at last appeared in print in 1687, Newton had done his major original work in science. It seems that he became bored with university life at Cambridge and welcomed an opportunity to move to the capital and find something new to do. Having gained some degree of public reputation by his opposition to King James II's ill-fated attempt to impose Catholicism on the University of Cambridge, Newton was elected a member of Parliament for the university's constituency in 1689 and again in 1701 (though he was defeated in 1703). He was then able to use his influence with those in power to gain for himself the position of warden of the Royal Mint in London, with the very large salary (for those times) of £2,000 per annum. He fulfilled his duties there with unexpected energy, pursuing counterfeiters zealously. In 1703, he became the first scientist to receive a knighthood, though it was apparently given more for his public services than his science. He thus attained a secure and very well rewarded position near the top of the English establishment of his day and was buried in Westminster Abbey. He was one scientist who made the most of his public fame.

Einstein's Attitude Toward Fame

When Einstein's general theory of relativity was confirmed in 1919 by observations that showed light "bending" near the sun during a solar eclipse, the media spread his name around the world, and he found himself the subject of more popular fame than any previous scientist. Hailed as the greatest scientific genius since Newton, Einstein found that his views were earnestly sought on every topic under the sun—politics, religion, and the meaning of life included. But none of this turned his head, as it might have done with a lesser man. He remained the same unassuming but unconventional, earnest yet humorous character that he had always been. He fit the stereotype of the eccentric and impractical professor, with his unruly shock of hair and his amateur efforts on the violin—and the world loved him for it. Enduring the attentions of the press and of politicians, cranks, help-seekers and hangers-on of all kinds, he would consent to be photographed only in return for a donation to the poor.

Thus, Einstein, though not seeking fame, was ready to make use of it. He was rather too kind in trying to help aspiring young scientists, with the result that possession of a glowing reference from Einstein became a *dis*advantage in the physics profession! He somewhat indiscriminately put his name and reputation behind various idealistic social and political movements. In the 1920s and early 1930s, he supported pacifism and conscientious objection to military service—for like many others, he had been horrified by the slaughter of World War I. He enthusiastically supported the ideals of the League of Nations in the 1920s and served on at least one of their committees. He also lent his name to the Zionist movement for the foundation of a Jewish state in Palestine. Einstein became increasingly identified with his Jewish heritage, if not with the religion. Later, he declined the offer of the presidency of Israel on the grounds of political incompetence and advancing age. In his last years, he lived almost as a recluse, avoiding the interviews and public involvements that the world continued to offer him. There was an important exception, for in the very last days of his life, he agreed to sign a document prepared by English philosopher Bertrand Russell, calling for an appraisal of the dangers of nuclear war. This led to the influential series of "Pugwash" conferences, which scientists from countries on both sides of the Cold War attended. Einstein viewed fame, once thrust upon him, as a means to promote causes he thought worthy.

Fleming's Enjoyment of Fame

Alexander Fleming attained extraordinary celebrity in the last decade of his life as the discoverer of penicillin, the first effective antibiotic. He traveled the world, collecting twenty-five honorary degrees, fifteen civic freedoms, and over one hundred and forty other miscellaneous honors and prizes. His fame even became extraterrestrial when a crater on the moon was named after him! But the main work on penicillin was done later by others (with whom Fleming eventually shared the Nobel Prize). So how did it come about that Fleming got the public fame? Did he seek it? Did he even deserve it?

As we have seen in 4.3, Fleming thought that penicillin would be useless as an antibiotic. After publishing the paper reporting his discovery, he turned his attention to other research. The effective development of penicillin as an antibiotic had to await the work of Chain and Florey in the early 1940s. Ernst Chain (1906–1979) was a German refugee biochemist who worked with Howard Florey (1898–1968), an Australian physician doing research at Oxford on Fleming's antibacterial lysozyme. There was heightened interest at the time in antibacterial agents because of the need to fight infections in wounds during World War II. Florey asked

Chain to try to find out the mechanism by which lysozyme kills bacteria. In checking through some two hundred papers for ideas, Chain came upon Fleming's 1929 report on penicillin, which suggested to him that the mold acted like the enzyme lysozyme. Florey and Chain obtained some cultures of the penicillium mold from Fleming's lab, which St. Mary's Hospital had been careful to maintain over the years. Florey's team then laboriously produced a few milligrams of purified penicillin from many gallons of liquid, having had to recruit hospital bedpans and factory tins into their "string and sealing wax" method of production.

The extract was then successfully tested on infections in mice. The first clinical trial was on a man with blood poisoning. He greatly improved after the first injection, but then they ran out of penicillin, and the unfortunate patient died—but not without having demonstrated penicillin's potential effectiveness. The British authorities, heavily occupied with the war effort, were not sufficiently impressed by the results of Florey's team to fund the production of more penicillin, so U.S. help was sought for mass production for wartime use. The U.S. government quickly came up with a million dollars for the project. Five large drug companies with hundreds of researchers worked to develop enough penicillin to serve the Allied troops. Within three years, there was a sufficient amount to treat millions of wounded people. This project was the largest single research venture ever undertaken up to that time.

At that dark stage of the war, the British government and press were desperate for good news of any kind, so it was natural for them to play up the story of the new miracle drug for all it was worth. Here was the "magic bullet" so long hoped for, which would would kill the germs without killing the patient. But why was it that Fleming got all the limelight, when the work to make the drug available for us had been done in Oxford by Chain and Florey? When the former head of Fleming's department heard of the success of penicillin, he wrote to *The Times* that Fleming should get the credit. Reporters flocked to interview Fleming. Then someone wrote in to suggest that Florey's group deserved the credit, and reporters went to interview him, too. But Florey imperiously turned them away, following the custom of many scientists of the time, especially in Britain, to shun publicity. (How many scientists could afford to do so nowadays?) So the reporters went back to Fleming, who was happy to talk with them.

Another factor in the story was institutional pressures and rivalries. St. Mary's Hospital was then a private foundation, dependent on publicity for its prestige and its income (this was before the advent of the British National Health Service, and the financial constraints more characteristic of the contemporary scene were at work even then). So the administrators of St. Mary's were happy to give the media the impression that a discovery of great benefit to humanity had been made within their doors. And the newspapers, especially those controlled by the patriotic Lord William Beaverbrook, embroidered the romantic story of the poor boy from rural Scotland who had become a scientific genius but remained modest and simple despite his fame.

Fleming submitted happily enough to what is now called "media hype," and his second wife avidly encouraged the process. However, he retained a realistic sense of his own part in the penicillin story, and he often paid generous tribute to the Oxford work. He said in 1945 that he did not really deserve his Nobel Prize. He would often say to his admirers, "I did not invent penicillin. Nature did that. I only discovered it by accident" (Macfarlane 1984, 260). Surely no one expected him to go around trying to refute all the false claims made about him by the press.

It is some measure of the modesty of his professional scientific standing that he was elected a fellow of the Royal Society only at the twelfth attempt and after attaining his public fame. Yet in his own undemonstrative way, he came rather to enjoy his worldwide celebrity. After all, he did not *have* to take on all the traveling and speechifying that he did. Fleming was received no less than five times by the Pope—an unusual fate for the lad from a land of the Protestant Reformation! At one degree ceremony in Spain, people in the street felt moved to kiss the hem of his gorgeous robe as he passed. Even in his native Scotland, a land less given to such emotional extravagance, a photograph shows a rare smile on his face as he is carried shoulder-high by the students of Edinburgh who had elected him as their rector.

It would be churlish to begrudge this very competent scientist his slice of luck and his hour of harmless celebrity at the end of a lifetime of undramatic laboratory work. But it would be unfair not to give at least equal mention of the part played by Florey, Chain, and others in producing penicillin. It is well to note that public fame is fickle and can be contrived by the media, not necessarily in proportion to scientific merit.

Suggestions for Further Reading

Broad, W. J., and N. Wade. 1982. *Betrayers of the Truth.*
Coulson, T. 1950. *Joseph Henry: His Life and Work.*
Desmond, A., and J. Moore. 1991. *Darwin: The Life of a Tormented Evolutionist.*
Harding, S. 1986. *The Science Question in Feminism.*
Hearnshaw, L. S. 1979. *Cyril Burt, Psychologist.*
Kamin, L. J. 1974. *The Science and Politics of I.Q.*
Macfarlane, G. 1984. *Alexander Fleming: The Man and the Myth.*
Sayre, A. 1975. *Rosalind Franklin and DNA.*
Watson, J. D. 1980. *The Double Helix* (Norton Critical Edition).

The Utility of Science

We speak of "utility" or "usefulness" as the motivation for scientific work that is undertaken for the sake of its likely practical applications. Francis Bacon argued that this should be the *primary* motive (see 2.1), but his claim has in fact more relevance to the science of the twentieth century than that of Bacon's own time. Historian of science Derek de Solla Price (who coined the term *high tech*) has asserted that basic science did not become significantly useful until the late eighteenth century, with the introduction of devices using electricity—culminating in the "invention factory" of Thomas Edison in America a century later (1985, 8). Practical technology largely preceded theory until then; it has been remarked, for example, that the steam engine did more for the science of thermodynamics than science did for the steam engine.

Some applications of science are of obvious human benefit, such as increasing food production, relieving pain, or curing disease. But in other cases, the intended application may be controversial, dangerous, or even evil in many peoples' eyes: for example, the manufacture of weapons of mass destruction (see 9.1) or the creation of new and sophisticated methods of torture or coercive control of behavior. A less clear example is the prospect of altering human genes, with its far-reaching promises and worries (see 9.2). If we describe the motive for militarily directed scientific work as "utility," we are saying only that some government or armed service saw such applications of science as serving their political and strategic purposes. New weapons are typically justified as necessary (even if regrettable) means for winning a war or for deterring potential enemies from using such weapons themselves.

At the commercial level, much scientific work is obviously motivated by its potential for profit to corporations in industrial applications. Research into how to give food a longer shelf life, for example, serves the ends of large companies that find it convenient and profitable to centralize food production in a few sites. Whether this is good for the consumer, in terms of taste, health, or variety of choice, is a different question. Research conducted by pharmaceutical companies may be motivated more by the realization that patents on their existing brands may be about to expire than by considerations of medical benefit. Obviously, what counts as "useful" depends on whom you ask and what purpose that person or institution has in mind. One person's use of science will often count as another's misuse.

6.1 The Tradition of Medical Science

There is a contrast to physical science, both in methodology and motivation, in the tradition of medical science. The practice of medicine as we understand it can be traced back to Hippocrates in ancient Greece, who set its goal as usefulness for human welfare. From the beginning, it was an empirically focused practice, unlike the one theoretical science besides mathematics that was well developed by the ancients, namely astronomy. The tradition of medical science only began to interact strongly with physical science when chemistry was applied to physiology in the nineteenth century. The science of medicine has by definition the goal of maintaining human health and well-being by the prevention and cure of illness and injury. This branch of science is most easily appreciated for its practical value by the public, and in recent years governments have tended to be especially generous in supporting medical research.

Medicine has always been at least as much an art (in the sense of a practical craft) as a science based on theory. But the practice of medicine with a scientific attitude has a long history of slow beginnings. The origins of medical science lie in a rather different kind of tradition than the physical sciences, however. The ancient Egyptian priestly class developed medicine extensively as part of its tradition. Unlike the Greeks, the Egyptians had no general prohibition on human dissection, and they conducted some basic surgery. But their practical knowledge was mixed with superstition and myths. Hippocrates (ca. 400 B.C.) is called the father of medicine, because he was the first individual in medical history who was not treated like a god. He made medicine a science by freeing the practice from its ties to the priesthood and its appeals to mysticism and supernatural influences. His famous aphorism *Ars longa, vita brevis* (The art is long, and life is short) expresses the idea of progressive accumulation of knowledge over generations. He emphasized the aim of disinterested service for human welfare, which is reflected in the Hippocratic oath, still taken by many medical students today.

The Hippocratic tradition was passed down in the writings of the medical school of Cos, dating from about 500 B.C. Lacking knowledge of human physiology and of biology and without instruments such as thermometers or microscopes, few scientific discoveries, in the modern sense, occurred during that era. But the Hippocratic approach to medicine was in the spirit of modern science. Its focus was on getting results and putting hypotheses to the empirical test. The treatise *On Ancient Medicine* protested against trying to base the study of medicine on cosmological speculation, advising its readers that "one must attend in medicine not primarily to plausible theories but to experience combined with reason . . . for a theory is a composite memory of things apprehended by sense-perception" (Farrington 1953, 75–76). The Hippocratic writing *On the Sacred Disease* contains an argument in favor of a naturistic account of what we now recognize as epilepsy:

> They who first referred this malady to the gods appear to me to have been just such persons as the conjurors, purificators, mountebanks, and charlatans now are, who give themselves out for being excessively religious, and as knowing more than other people. Such persons, then, using the divinity as a pretext and screen of their own inability to afford any assistance, have called the disease sacred. . . . If they imitate a goat, or grind their teeth, or if their right side be convulsed, they say that the mother of the gods is the cause. But if they speak in a sharper and more intense tone, they resemble this state to a horse, and say that Poseidon [god of the sea] is the cause. . . . But this disease seems to me to be no more divine than the others; but it has its nature such as other diseases have, and a cause whence it originates. (Hippocrates 1952, 154–155)

The Hippocratic writings begin to express what we would call the scientific approach to understanding disease.

Harvey: The Circulation of the Blood

Every schoolchild knows that the blood circulates because of the pumping action of the heart. Yet an accurate description of the heart's function and the very idea that blood circulates through the body required a breakthrough in medical science. This came in 1628 with the publication of *An Anatomical Treatise on the Movement of the Heart and Blood* by William Harvey (1578–1657).

Previous ideas about this fundamental element of physiology were derived from the works of Greek physician Galen (A.D. 129–199), which had long been considered authoritative on human anatomy. He proposed that the blood makes the heart move, rather than the heart actively pumping the blood. Galen's authority was first challenged by Vesalius (1514–1564), who set the foundation for modern medicine in his masterwork *On the Structure of the Human Body*. This text was printed in 1543, the same year in which Copernicus's revolutionary work was published. Its accurate and artistic depiction of human anatomy is not much different from what you would see in works used by medical students today. Vesalius had begun his studies at the University at Padua in Italy, where the medical tradition of antiquity was still flourishing. At that time, Galen's work was still unchallenged, despite a large number of inaccuracies and misconceptions in them. Following Aristotle's example, Galen had based his study of human anatomy largely on dissections of Barbary apes rather than of actual humans. Human dissection was not considered proper in the Greek tradition and had only become accepted—using the corpses of criminals—in a few places in Renaissance Europe.

Unlike the other professors of anatomy, Vesalius did his own dissecting, instead of reading from Galen while barber-surgeon technicians did the actual work. In 1555, Vesalius announced that the pores believed by Galen to allow blood to pass from the right to the left chambers of the heart simply did not exist. Bold though he was, Vesalius found it difficult to muster enough courage to challenge the traditional authority. He wrote: "Not long ago I would not have dared to diverge a hair's breadth from Galen's opinion. But the septum [dividing wall between the chambers of the heart] is as thick, dense and compact as the rest of the heart. I do not see therefore, how even the smallest particle can be transferred from the left to the right ventricle through it" (Mason 1962, 216–217).

Even so, Harvey was the first openly to reject the cardiac pore theory. After all, it had not always been safe to challenge established authority on such matters. Michael Servetus (1511–1553), with whom Vesalius had worked in Paris, was the first to suggest that the blood passes from the right to the left chamber of the heart by way of the lungs. Servetus had proposed that the blood is really one substance, in contrast to the commonly held belief that the red arterial blood was essentially different from the venous blood. His Unitarian theology gave him the idea of unifying Galen's three bodily "spirits": the natural spirit of venous blood, the vital spirit of arterial blood, and the animal spirit of the nervous system. He believed that the divine spirit in the air (what we call oxygen), purifies the venous blood, making it red. Servetus anticipated Harvey's view that arterial blood is the same fluid as the venous blood except for the change that takes place in the lungs. Servetus's reward for such theorizing was to be burnt at the stake by John Calvin, the leader of Protestantism in Geneva. What Calvin considered his worst heresy was the claim that the soul amounts to a spiritual essence in the blood.

Harvey studied in Padua under Girolamo Fabrizio, a pupil of Vesalius, from whom he learned a curious fact about veins: They have valves. Harvey noted that these valves ("little doors") are arranged so that the blood flows in one direction only, toward the heart. Harvey cited this fact when Robert Boyle (1627–1691) asked him what had led him to think of circulation of the blood. Harvey replied that he couldn't conceive of such a device as the valves in veins being present without having been designed for some function. He spent twenty years studying the beating of hearts in animals, from frogs to dogs, to test his hypothesis of circulation. In the spirit of the awakening empirical approach to nature, Harvey inquired what mechanisms operated in the body. His suspicion that the blood makes complete cycles through the body was motivated, curiously, by the one example Aristotle cited of perfect circular motion among earthly phenomena. Harvey wrote that he "began to think whether there might not be a motion as it were in a circle, in the same way as Aristotle says that the air and the rain emulate the circular motion of superior bodies" (Mason 1962, 221).

Harvey made numerous experiments trying to measure the volume of blood flow. He found that within half an hour, more blood flows through the heart than exists in the entire body—one of the first applications of quantitative reasoning in biology. He made what seems the obvious inference: The blood must in some way circulate, since it could hardly be created and destroyed in such a short time. He hypothesized that the heart acts like a pump, using mechanical analogies that were becoming current at the time. He observed that the hearts of animals could actually be felt to harden as they contracted, just like other muscles do when they contract. He even tested directly Galen's claim that blood passes through pores in the heart wall by cutting away the left ventricle of the heart of a dog. He saw no flow from the right ventricle. He was, however, unable to fully trace the circulation of the blood from the fine arteries back to the veins. It was left to Italian physiologist Marcello Malpighi (1628–1694), using a microscope, to discover the tiny capillaries that link arteries to veins. Harvey's discovery raised questions about what happened to the blood in the lungs, where the "vital spirit" was added, but medical science had to wait nearly one hundred and fifty years for chemistry to discover the role of oxygen (see 4.3 on Priestley).

Pasteur: Controlling Microorganisms

Louis Pasteur (1822–1895) is the father of modern microbiology. He provides a striking example of a man with enormous gifts and plenty of inclination to devote them to pure science, spending much of his effort on extremely practical problems in brewing, agriculture, and medicine. He has become, quite literally, a household name, in that much of the milk we consume has been "pasteurized." But Pasteur's contributions are much wider and deeper than that one well-known process; he can claim much of the credit for propounding and verifying the germ theory of disease.

After gaining admission to study at the École Normale Superieure in Paris, Pasteur specialized in physics and chemistry, going on to do doctoral research in crystallography. In this field, he soon made the striking discovery that the crystals of certain substances come in two forms, displaying left-handed and right-handed asymmetries. This early research provided the basis of modern stereochemistry, the study of the spatial configuration of atoms. As professor of chemistry at the University of Strasbourg, Pasteur continued to study such chemical asymmetries and speculated that they might be distinctive of living matter. At the end of his life, he was to express some regret that he had not pursued this pure research further.

Pasteur's next professional move, to take up the chair of chemistry at the newly organized University of Lille in 1854, seems to have been motivated by the prospect of better experimental facilities. But part of the deal was that he should focus his teaching and research on topics of interest to the industries of the Lille region, such as wine making. Here is a nineteenth-century example of a very contemporary theme—funding being given only on the condition that the scientist contributes to the purposes of the funding agency. Pasteur seems to have found this a stimulus rather than a burden. Until then, he had been a theoretical scientist doing research for its own interest, but for the rest of his life, he devoted most of his efforts to solving practical problems. He told his students that he saw no conflict between pure and applied science and that the two interact so much that there are not really two kinds of science but instead only science and its applications. In conformity with this philosophy, Pasteur was soon contributing to the improvement of alcoholic fermentation, a process dear to the hearts and the pockets of the people of Lille. Some years later, after the Franco-Prussian war, he patriotically proposed to make French beer better than its German rival!

A short paper Pasteur wrote in 1857 on the fermentation of milk marked the beginning of scientific microbiology, for in it he postulated the role of living things in chemical change. When milk goes sour, the milk sugar (lactose) is transformed into lactic acid. He showed that this process depends on the presence of huge numbers of microscopic creatures that can reproduce themselves rapidly as long as sugar is supplied (we now call these tiny living organisms yeasts, microbes, or bacteria). Generalizing from this discovery, he put forward the theory that other natural chemical processes like fermentation, putrefaction, and so on are each dependent on a specific kind of microorganism. It was not previously known that this was the true nature of the process of making alcohol, which had been carried on for thousands of years on a trial-and-error basis, with its results being among the most-appreciated gifts of "civilization"!

In 1857, Pasteur returned to Paris to assume a senior position at the École Normale, where he had heavy administrative duties and ample opportunity to indulge his tendency to get involved in public controversies. Despite these constraints on his time and the scanty facilities that the École offered him, his laboratory research went ahead. He developed the techniques of aseptic manipulation and sterilization, which are standard procedure today. He took out patents on some of these methods and used the proceeds to fund further research as well as to provide for his family. But on the whole, his scientific career seems to have been a case of genuine dedication to public welfare, with the desire for reward present but not primary.

The other great achievement of Pasteur's life was the germ theory of disease. Pasteur's experiments, along with the later work of German microbiologist Robert Koch (1843–1910), demonstrated conclusively that germs could cause diseases in higher animals and humans. In this case, too, the research that was to prove so important started from an economic problem. Soon after arriving at his explanation of fermentation, Pasteur had conjectured that each infectious disease might be caused by a specific kind of microbe. He was given an opportunity to test this idea when the minister of agriculture asked him to head a commission to inquire into a disease that was infecting the silkworms in central France and ruining the silk industry. Pasteur protested that he was a chemist, not a zoologist, but the minister insisted, and Pasteur's desire to be useful to his country seems to have induced him to take up the challenge. From 1865 to 1870, he devoted himself to studying the diseases of silkworms, working for much of that time in a makeshift laboratory in the Cevennes mountains. Despite the death of his daughters and a cerebral hemorrhage, his indomitable will kept him going.

Pasteur's first experiments to develop vaccinations to prevent disease focused on anthrax, chicken cholera, and rabies (which is caused by a virus, not a bacterium), using a method already known to work for smallpox. The theory was that injecting a weakened or "attenuated" strain of the relevant microbes could activate the natural defense systems of the body, thus preventing subsequent infection without inducing a full-blown case of the disease. To combat skepticism, Pasteur arranged dramatic public confirmations of the effectiveness of immunizing animals against anthrax. In 1885, he achieved his most-remembered success in saving a boy from rabies, which until then was 100 percent fatal. A young boy appeared at his door, having been bitten by a rabid dog and already showing some of the symptoms of the dreaded hydrophobia, as the disease was then known. Pasteur had the courage to try the first human vaccination against rabies. It proved successful; the boy went on to live fifty-five more years, and Pasteur went on to enjoy fame, honors, and rewards, including a national pension from the French government.

However, Pasteur has been recently accused of unethical behavior not fitting his folk-hero image of a selfless benefactor of humanity. He kept detailed notebooks of his research but told his family to keep them secret, and they were not made public until 1971, when his grandson died. Historian Gerald Geison has recently studied them and found serious discrepancies with Pasteur's public versions of his work that give grounds for several accusations of scientific and clinical misconduct. For one, Pasteur's published account of his use of a rabies vaccine on the boy he saved was misleading; he used a rather different procedure than the one he had been successfully testing on dogs. The criticism is that he could not have known at the time whether his treatment was safe and effective. But since the boy would surely have died without treatment, Pasteur's action can be justified by saying that there was nothing to lose by trying an experimental treatment in such a case. The criticism remains that Pasteur's published account was misleading about the exact nature of what he had done. It seems that in his efforts to convince scientists of the truth of his theories and to persuade the public of the efficacy of his methods, he was prepared to be less than fully honest about the details of his procedures.

This tendency of Pasteur's is confirmed by another instance in which he has to be accused of fraudulent presentation, however right he turned out to be in scientific substance. In his public demonstration of a vaccine against anthrax in sheep in 1881, he gave the impression that he was using his own oxygen-attenuated vaccine, but his notebooks suggest that he used a different one prepared by a collaborator. He was either lucky or had very good scientific intuition, for a few months later he was able to perfect the oxygen-attenuated vaccine, which then became standard. There is, however, a darker side to this case. The procedure he used had been developed earlier by a veterinarian named Toussaint, who had told Pasteur about it. Pasteur took the glory, but Toussaint had a nervous breakdown and died.

Besides being a supreme experimentalist, Pasteur was a formidable adversary in controversy, an adroit player of academic politics, and an effective crusader to get his theories applied in practice. Wherever he saw human benefits to be gained, he set out to persuade his doubting or ignorant contemporaries that his proposals would work in practice. He competed for many awards and prizes from private societies and governments. "Successful" public demonstrations were politically and financially crucial for him, which explains, but hardly excuses, his departures from the truth. He curried favor with those in power, including the Emperor Louis Napoleon and his court. Toward the end of his life, his public status was such that he was induced to run for election (as a conservative) to the French Senate, presenting himself as "the candidate of science and patriotism," but he was roundly defeated. He is an example of scientific expertise, curiosity, and determination applied to benefit humanity but mixed with other familiar but less admirable human traits, such as egotism and desire for fame, power, and money.

Salk: Fame and Controversy from a Polio Vaccine

Jonas Salk became one of the best-known scientists in the United States when in 1954 he came up with a vaccine against the dreaded poliomyelitis (polio), or infantile paralysis. Polio has now been virtually eradicated by use of the Salk vaccine and a later one introduced by Albert Sabin. Behind the success story, however, is a complex and still controversial account of how science and its application to public health should be conducted. Let us first consider some of the history of vaccination.

Country doctor Edward Jenner (1749–1823) introduced the first systematic use of vaccination to prevent disease, a medical innovation that has probably saved more lives than all medical drugs combined. In 1798, he demonstrated that inoculation of material from cowpox blisters could immunize against smallpox, a disease that was killing up to a fourth of the population of England in the seventeenth century. Jenner's work is a striking case of practical knowledge preceding theoretical understanding. In his time, there was no concept of microorganisms causing disease. Microscopist Anton van Leeuwenhoek (1632–1723) had actually observed microorganisms (which he called "animalcules"), but it wasn't until almost a century after Jenner that Pasteur developed the germ theory of disease.

In 1718, Lady Mary Wortley Montagu (1689–1762) introduced inoculation to England from Constantinople (Istanbul), having gotten the idea from Turkish and Arab women who had been using it for centuries— the technique originated in China. Individuals exposed to infection from mild varieties of smallpox became resistant to more serious forms. Lady Montagu had her own children inoculated and on returning to England, she urged the Prince of Wales to have his children inoculated, too. He did so, successfully, but only after testing the procedure on a group of criminals and orphans. Nearly a century later, Jenner inferred from the similarity between cowpox and smallpox that the body's defenses might be stimulated by the milder disease to give immunity from the more virulent one. Using a thorn, he inoculated a boy with material from an infected milkmaid. Six weeks later, the boy was exposed to smallpox but was unaffected. In 1798, Jenner published an account of his observations, entitled *Inquiry into the Cause and Effects of Variolae Vaccinae.* (The term *vaccinae* refers to cowpox, from the Latin *vaca,* "cow.") Several hundred people who had been vaccinated came down with smallpox, but Jenner was able to show that the serum used was not properly prepared. At first, there was much controversy over the safety and effectiveness of Jenner's vaccination, but he lived to be acclaimed a hero. He was knighted and rewarded with £20,000 from Parliament, and he received an honorary degree from Oxford, letters of praise from Napoleon and Jefferson, and a gold ring from the czar of Russia. Both the controversy and the fame were to occur again for Salk with his polio vaccine. Although Jenner was not the first to inoculate against smallpox, he deserves credit for developing and gaining acceptance for an effective vaccination.

The smallpox virus, as we now understand it, becomes so attenuated (weakened) by transmission through infection of a cow that it causes only a mild reaction in humans, as opposed to the often deadly and disfiguring smallpox. The idea was generalized into a theory of immunization against diseases: Doses of attenuated infectious material would stimulate the body to fend off the same infectious agents later on. By the 1880s, Pasteur had introduced inoculation to give immunity against a large range of diseases. With the germ theory of disease, the mechanisms by which immunization works began to be understood, although a detailed understanding has emerged only in recent years. Immunology is a large research area today, with attempts to understand autoimmune diseases such as arthritis and to develop immunizations, especially against the HIV (AIDS) virus, which attacks the body's very immune defenses.

Polio did not begin to stand out as a serious illness until the twentieth century, partly because many of the diseases that previously had taken so great a toll had been reduced, by vaccination in the case of smallpox and by improved sanitation in the case of cholera. In fact, the incidence of death and paralysis from polio actually increased because of better sanitation. Polio is usually a relatively mild infection of the intestines. Most babies are born with some temporary immunity and if exposed to the virus while still immune, they develop a more permanent immunity. Thus, sheltering children from unsanitary conditions early in life increased the chances of their later suffering more severe infections of polio.

Although the number of deaths from polio never approached those from other epidemics, the effects of paralysis, striking most commonly children, made the disease especially terrifying. President Franklin D. Roosevelt was a famous victim who was crippled by the disease as an adult, after which he was essentially confined to a wheelchair. Roosevelt's going on to become president helped raised public interest in polio. In 1938, the National Foundation for Infantile Paralysis was organized, and funds were raised by an annual and widely publicized March of Dimes campaign. In the initial appeal, the public sent in several million dimes to the White House, thus making the campaign the largest voluntarily funded research program in medical history. Only after a successful vaccine was developed did the U.S. government step in. The development of polio vaccines marks a transition from private medical research to the largely governmental funding of the big medical research programs today.

Jonas Salk enrolled in the City College of New York at age fifteen and went on to be a top student in medical school. Although he was warned that it was not the way to make money, he chose to go into research and spent an extra year studying biochemistry after getting his M.D. While in medical school, he met Dr. Maurice Brodie, who was doing research on polio, and Dr. Thomas Francis, with whom he worked on a vaccine against influenza, using viruses killed by ultraviolet light. In 1935, Brodie had tried out a vaccine against polio using viruses killed by formaldehyde. Hundreds of children were inoculated, and hopes were raised, only to have the experiment end in dismal failure. Some of the children came down with polio caused by the vaccination itself, and it did not grant effective immunity. More basic research was needed.

As we have seen, Jenner's vaccination against smallpox was successful, but not because he had any understanding of the underlying cause of the disease (a virus) or of how vaccination confers immunity. Research to develop a polio vaccine also began with very little knowledge of the virus. (Jenner had introduced the term *virus*—"poison" in Latin— to label the unknown agent causing smallpox.) Even by the 1930s, the viruses hypothesized in 1908 by Viennese researcher Karl Landsteiner to cause polio were difficult to work with since they could not be seen with optical microscopes and were difficult to culture. Research on polio served to stimulate work in the field of virology, which later contributed to the momentous discovery of the structure of DNA. It was thought that the polio virus would only grow in live nerve tissue. But if the virus grew only in nerve cells, prospects for a safe vaccine looked bleak. In 1949, a team led by John Enders succeeded in growing polio viruses in nonnervous human and monkey tissue. In 1952, Dorothy Horstmann and David

Bodian independently made a second breakthrough by showing, against the dominant belief of virologists at the time, that polio viruses did pass into the bloodstream. It was now established that paralytic polio starts in the intestines, passes into the blood, and then invades the nerves.

The fact that there are three basic types of polio virus, with different "strains," or subtypes, within the types, made Salk's study especially difficult. Salk had to begin his polio research with the tedious job of distinguishing the types of polio virus in a complex process requiring the sacrifice of some 17,000 monkeys. The lack of knowledge of these different types was the primary reason for Brodie's failure to develop a safe and effective vaccine in 1935. Research just to determine how many strains of virus cause polio cost over a million dollars. Mass publicity was necessary to raise the funds since little support came from the government until after the successful trials of the Salk vaccine. But publicity in the development stage tended to interfere with the scientific research, as it introduced considerations beyond questions of scientific theory and experiment.

Salk's work on a polio vaccine began in 1947 at the University of Pittsburgh, sponsored by the March of Dimes Foundation. He decided to go for a killed-virus vaccine. Effective killed-bacteria vaccines had already been developed, so he did not hold to the dogma that vaccines must use live viruses. One of Salk's discoveries, one that many prominent virologists found hard to accept, was that the capacity of viruses to induce antigens did not correspond to their infectivity. He reasoned that a weakened live-virus vaccine might be less safe since the rapid mutation of polio viruses might make them virulent again. In 1949, Dr. Isabel Morgan at Johns Hopkins had succeeded in immunizing monkeys against polio using a killed-virus vaccine. To get a safe and effective vaccine from dead viruses required making sure that the viruses were no longer infectious yet retained their capacity to stimulate the body to make antibodies. Salk proceeded patiently by trial and error to produce a vaccine. The procedure required using a precise amount of formalin, for just the right amount of time, at just the right temperature and acidity. He successfully tested his vaccine on monkeys, but uncertainties remained about how it would work with humans. Salk first tried his vaccine on children who had already suffered paralytic polio, reasoning that the vaccine could do them no harm and yet would provide a test as to whether the vaccine raised levels of antibodies in the blood.

Until 1953, Salk's work was not publicized outside the scientific community, but it was a different game entirely when rumors of a polio vaccine reached the news media. An article had been prepared for the *Journal of the American Medical Association* to announce plans for a public test, but word of the promising new development leaked to gossip columnist Earl Wilson, who wrote a newspaper article entitled "New Polio Vaccine—Big Hopes Seen." This led to a public clamor for the vaccine and criticism of Salk as a publicity seeker, though he had nothing to do with the newspaper story. To clarify the status of the vaccine, Salk spoke on national radio and TV in March 1953. He explained the basic principle of vaccination in simple terms:

> A vaccine is made of the virus that causes the disease. Then, when the vaccine is injected, the body reacts with the formation of antibodies. These antibodies are found in the blood and remain to defend against future attacks. It is evident, of course, that the virus contained in the vaccine must be rendered harmless so that when injected it will not cause disease but will result only in the formation of projective antibodies. (quoted in Carter 1965, 160–161)

But he had to announce that despite recent progress, there would not be sufficient vaccine available for widespread use by summer 1953.

With the public now aware of Salk's work, pressure built for getting the vaccine out before the 1954 polio season. There was popular demand for immediate vaccination, but Salk insisted on further testing to confirm that the vaccine was both effective and safe. In the tradition begun by Jenner, he inoculated himself and his family. Several hundred volunteers took the vaccine and it proved successful in all cases. The massive medical experiment began in April 1954, using the vaccine on nearly half a million schoolchildren in the first through third grades; a couple hundred thousand more were injected with a placebo for control. Several large drug companies began to manufacture the Salk vaccine in bulk, allowing for a large-scale test in 1955, when nearly two million school children were inoculated. Vaccination proved nearly always effective in preventing polio.

When a report of the vaccine's success was presented, church bells rang and sirens sounded to acclaim this miracle of modern medicine. Salk was greeted with an outpouring of praise and thanks for conquering the dreaded polio. He became an international hero. A Jonas Salk dime was suggested in Congress; magazines were full of articles about him; and movie studios considered doing Salk's life. But unlike Fleming (see 5.3), who not only received public acclaim but a Nobel Prize, Salk received little recognition from the scientific community. His name was raised at Nobel Committee meetings, but he lacked recommendations from eminent virologists. He even failed to be admitted to the National Academy of Sciences, partly, it has been suggested, because of his great public fame.

Salk seems not to have desired the clamor associated with public fame, especially at a time when it might interfere with his scientific work. After the public associated his name with the polio vaccine, he had little peace. He wondered, "Why do they all want to know what I eat for breakfast?" (Carter 1965, 166). Commenting on his own motivation, he said:

> I wanted to be a scientist, not a vaccine manufacturer. I was an immunologist, not a product developer. . . . As we now know, the selection of one of our experimental vaccines for commercial manufacture and field testing subsequently became a matter of urgent, humanitarian concern. . . . This transformation of a scientific worker into a product developer was most unusual. It happened because of the unique nature of the given situation, but was not planned or plotted or looked forward to, nor could it have been. (quoted in Carter 1965, 67)

The actual course of the vaccination program was not at all smooth. Much debate ensued about whether the testing had used double-blind studies, a common scientific procedure to ensure greater reliability that could, however, pose practical and ethical problems. Salk initially wanted to be able to revaccinate if a certain batch of vaccine showed unsatisfactory results. And he was concerned about giving placebo shots to children who would then receive no immunity against coming down with paralytic polio. In an emotional letter to the president of the National Foundation for Infantile Paralysis, Salk even questioned the use of placebo controls as "a fetish of orthodoxy" that "would serve to create a beautiful epidemiologic experiment. . . . but would make the humanitarian shudder and would make Hippocrates turn over in his grave" and attacked what he considered "false pride based upon values in which the worship of science involves the sacrifice of humanitarian principles on the altar of rigid methodology" (Carter 1965, 192).

In April 1954, shortly before the field test, foes of Salk's project enlisted the aid of another Broadway columnist, Walter Winchell, who announced:

> In a few moments I will report on a new polio vaccine—it may be a killer! . . . The National Foundation for Infantile Paralysis plans to inoculate one million children . . . with a new vaccine this month. . . . The U.S. Public Health Service tested ten batches of this new vaccine [and] found (I am told) that seven out the ten contained live (not dead) polio virus. . . . The name of the vaccine is the Salk vaccine. (Carter 1965, 232)

This led many parents to withdraw from the 1954 vaccination test. Despite the overall success, one defective batch of vaccine produced by Cutter Laboratories almost undid the whole program. Cutter's vaccine, made without all of Salk's stringent controls, caused at least two hundred and fifty cases of paralysis and eleven deaths. Despite the lack of problems among most of the five million children vaccinated with a commercial vaccine after the successful field trial, the surgeon general announced in May 1954 that vaccinations would be suspended until safety checks could be made on all vaccine supplies. The vaccines produced by Parke, Davis, and Eli Lilly were, however, released as safe a few days later. This suspension caused considerable confusion—parents feared both having their children vaccinated and not having them vaccinated.

There had always been much competition in the race to develop a polio vaccine, and Salk's version, despite its general success, was relentlessly attacked by virologists who believed that only an attenuated live-virus vaccine could be safe and effective. Salk's archfoe was Albert Sabin, who fought unceasingly for a live vaccine. Sabin's research in 1935 indicated that polio virus only grows in nervous tissue, but this conclusion depended on use of an atypical strain. If he had tried other strains, he might have been the first to develop a polio vaccine. Sabin continually expressed doubts about Salk's project. In June 1953, less than a year before Salk's successful field test, Sabin stated flatly before an AMA (American Medical Association) audience that "a practicable vaccine for poliomyelitis . . . is not now at hand." He went on to express the objective of polio research as he saw it:

> To imitate what nature does for 99 to 99.9 percent of the population but without incurring the one in a hundred to one in a thousand risk of paralysis which in many parts of the world is the price for acquiring immunity to poliomyelitis. . . . the ultimate goal for the prevention of poliomyelitis is immunization with "living" avirulent virus which will confer immunity for many years or for life. (quoted in Carter 1965, 179–180)

Salk rejected the idea that a "natural" vaccine must be superior to one using an artificial killed virus. He sarcastically referred to his own "heresy" as the belief that a killed vaccine could "contain the magical life force of the natural disease" (196).

Sabin lost his initial fight against the widespread use of the Salk vaccine but later triumphed, at least with the medical establishment in the United States. In 1959, Sabin's live-virus vaccine was tested on millions in Russia and Eastern Europe. Though there was some mistrust of the results of testing in communist countries, by 1960 Sabin's vaccine had become widely accepted, and it was licensed for use in the United States in 1961. The live-virus vaccine was easier to administer (orally in sugar cubes) and was supposed to be safer and to confer longer-lasting immunity. But for Sabin "the church bells did not ring. . . . No proposals were made for Sabin dimes. There was no presidential citation or any of the rest of the canonization syndrome that had greeted Salk. The people had grown weary of polio heroes" (Klein 1972, 147).

Sabin's vaccine never had a disaster like the defective Cutter batch of the Salk vaccine. But by June 1964, there were 123 cases of paralytic polio "associated" with the Sabin vaccine—the paralysis had appeared within 30 days of receiving it. Sabin's vaccine gained the support of the U.S. government and most of the medical establishment despite the fact that Sabin's vaccine continued to cause some cases of paralytic polio, whereas Salk's vaccine, where used exclusively as in Scandinavia, resulted in total eradication of the disease. Salk and his supporters charged that the decision to use Sabin's vaccine was "unreasonable" and based on commercial and political considerations rather than on scientific evidence. The controversy over the relative advantages of Salk's killed-virus vaccine versus Sabin's live-virus vaccine continues to this day.

6.2 Practical Technology and Science

Watt: Science and the Steam Engine

A famous example of a technological innovation that was directed toward practical utility but also stimulated the development of theoretical science is the steam engine. James Watt (1736–1819) is often described as the inventor of the steam engine, although what he actually did was to effect a crucial improvement on previous designs. He was more of an engineer than a scientist by our definition (see 3.1), but the development of efficient power technology was a crucial factor in the Industrial Revolution, and it marked the beginning of greater interaction between scientists and inventor-engineers.

Born near Glasgow in Scotland, Watt displayed technical aptitude early in life, making models in his father's workshop. Since a university education was financially out of the question, he planned to earn his living by designing and constructing mathematical instruments. After a year's training in London, Watt became an instrument maker for Glasgow University. This was not a paid position, but he was permitted to set up a shop within the university, and thus he was already in business on his own account (an early example of "enterprise culture" in an academic institution!). He was not a purely profit-seeking entrepreneur, however, for when he invented an ingenious apparatus for drawing in perspective, he failed to patent it, and his London competitors copied it.

An early form of steam engine had already been developed by Newcomen. When Watt was asked, in the course of his work at Glasgow, to do some repairs on a model of this form of engine, he soon realized that the reason it consumed so much steam was that the cylinder was heated and then cooled down again with every stroke of the engine. Crucial to this insight were certain elements of very recent scientific theory that he had learned from Joseph Black, the professor of chemistry at Glasgow University. In particular, the concept of latent heat implied that it takes a certain amount of heat input to convert water into steam, even though the temperature of the water is not thereby raised. The solution to the engineering problem—said to have occurred to Watt during a Sunday walk on Glasgow Green—was to add a separate chamber in which the steam could be condensed without cooling the main cylinder.

The route from technical ideas to industrial application was far from smooth, however. Watt immediately began trying to construct models in his workshop, a long, frustrating process of trial and error. He had no capital resources to set up full-scale engines, and Black put him in touch with a Scottish industrialist named Roebuck, who went into partnership with Watt. Financial success did not smile on them, and Roebuck went bankrupt in 1775. Watt was then forced to earn his living in other ways, and for years he was employed on land surveying for canals in Scotland. The industrial production of Watt's improved steam engine was eventually undertaken by Matthew Boulton of Birmingham, who formed with Watt one of the most successful business partnerships of all time. Boulton had the capital, commercial sense, and self-confidence, and Watt supplied technical expertise and fertility of invention. The new steam engine was patented, and Boulton and Watt eventually became rich from it. The royalties were figured as a percentage of the cost of fuel saved in comparison with previous engines, which required exact calculations of their output and efficiency. To this end, Watt invented a variety of measuring devices and introduced *horsepower* as a unit of measure. With a clear eye to their profits, Boulton and Watt resisted the development by others of the newer high-pressure form of steam engine as long as their patents were valid (until 1800). They had to take legal action to defend their existing patent against infringement. During a visit to Paris, French chemist Claude Louis Berthollet (1748–1822) told Watt of a bleaching process involving chlorine. When Watt suggested that the process should be patented for its commercial value in the textile industry, Berthollet is said to have replied, "When one loves the sciences, one doesn't need a fortune" (Gillispie 1976, vol. 14, 197). As we shall see in Chapter 7, the profit motive does not always sit easily alongside pure scientific curiosity.

Watt's forte was making useful inventions, but he did have some interest in science itself. At Glasgow, he had learned some chemistry from

Black, for whom he performed experiments in his early days. Though he never became an academic scientist doing basic research, he joined the Lunar Society, the most famous of a number of provincial associations of scientists, inventors, doctors, and businessmen that flourished in England at the time of the Industrial Revolution. His interest in chemistry was rekindled by the work of Priestley (see 4.3), who was also a member of the Lunar Society. In 1783, Watt made the prescient suggestion that water was not an element but a compound. (Lavoisier gets the main credit for this, since he actually specified the composition of water as H_2O.) We can plausibly speculate that if circumstances had been different, if a university education had been possible for him, Watt could have become much more of a scientist, instead of or in addition to being a great inventor-engineer. But social circumstances bent his talents and efforts in the direction of usefulness to industry.

Watt's invention of a more efficient steam engine led to more fundamental research in the theory of heat (see the discussion of Rumford in the following section and of Gibbs in 4.1). The science of thermodynamics, with its general equations for the transformation of energy, originated from attempts to understand how to get power from steam engines even more efficiently. In the 1830s, French physicist Sadi Carnot (1796–1832), interested in improving the efficiency of Watt's engine, developed a theory for calculating the maximum efficiency of converting heat into work. Thus, even though Watt himself did not contribute to scientific theory per se, his work did focus attention on problems that gave rise to new physical theory.

Rumford: Science for Human Welfare

Sir Benjamin Thompson (1753–1814), who became Count Rumford, was an American-born physicist and inventor. He had a rather more colorful career than most scientists. Born in Concord (originally called Rumford), Massachusetts, he married a rich widow of thirty when he was eighteen, fought with the British during the American Revolution, became a member of the Royal Society at age twenty-two, was a member of both the French and American Academies of Science, married the widow of Lavoisier (who was herself something of a scientist), and was made a count of the Holy Roman Empire for his services to the king of Bavaria. Clearly the sources of funding for his research were many and various! The inscription on his gravestone near Paris marks him as a "celebrated natural philosopher, renowned philanthropist" and remarks on "his discoveries in light and heat," concluding that "his works to improve the lot of the poor will make him always cherished by the friends of humanity" (Bradley 1967, 2).

Rumford is usually remembered in textbooks for inferring the equivalence between mechanical energy and heat from the well-known phenomenon of cannons becoming hot when fired. He measured for the first time, with fair accuracy, the relation between heat and work. In 1798, he published a paper arguing against the then popular belief that heat was a special substance called "caloric." He thus anticipated the kinetic theory of heat, which was not accepted by the scientific community until nearly a century later. But he spent much of his time on applied science and technology. Living up to the inscription on his grave, he was a prolific inventor of devices that have contributed greatly to more comfortable living. Among his inventions are the first effective photometer, an efficient fireplace, the sofa bed, and the drip coffee maker!

Count Rumford was highly ambitious. He gained a reputation as a careerist and adventurer, in some respects anticipating contemporary scientists who travel around gathering honors and giving advice to governments and industry. He became wealthy and rather famous in his time, but he never took a cent for any of his inventions. Working for the king of Bavaria, he reformed the army and experimented to find more effective insulating clothing for the soldiers. In another of his projects, Rumford experimented with animals to test ideas on nutrition and concocted what were called "Rumford soups" that became famous throughout Europe. He was far ahead of his time in applying systematic social planning, developing, for example, the first significant public works program in Bavaria, solving the problem of the numerous beggars and thieves roaming the streets. A statue in Munich proclaims his idea that "to make vicious and abandoned people happy it has generally been supposed necessary first to make them virtuous. But why not reverse the order? Why not make them happy and then virtuous?"

One of Rumford's most important contributions to the development of science was founding the Royal Institution of London, dedicated to "bringing forward into general use new inventions, and improvements, particularly such as related to the management of heat and the saving of fuel and to various other mechanical contrivances, by which domestic comfort and economy can be promoted." Like the older Royal Society (see 1.2), the Royal Institute also supported theoretical science. It provided the laboratories in which chemist Sir Humphrey Davy (1778–1829) and his protégé, the great experimenter Michael Faraday (1791–1867), made their important discoveries and still carries on the tradition of public discussion of science today.

Carver: Scientific Agriculture

George Washington Carver (1864–1943) is often not included in biographies of scientists, but he has some importance as an innovator in the application of scientific thinking to practical problems in agriculture. Born a slave in Missouri toward the end of the Civil War, Carver rose to become something of a folk hero by applying science to agriculture for the betterment of poor black farmers in the American South. He and his mother were abducted from the Carver family by slave rustlers when he was only a few months old. His mother was never heard of again, but his owner, Moses Carver, got the infant George back by trading a $300 racehorse for him. The childless Carvers adopted him and his brother after the slaves were freed in 1865.

Carver was not allowed to attend the local all-white school, but he did go to an elementary school for blacks in another town. When he was about fourteen, he traveled to the Midwest, taking odd jobs and always seeking to further his education. When he was twenty, he tried to enroll in a small Presbyterian college in Kansas. He was accepted by mail, but upon arriving at the college, he discovered that it did not admit blacks as students. This discouraged him from seeking further education for several years. In 1890, he was encouraged to apply to Simpson College in Iowa to study art, having shown promise as a painter. Though lacking a high school diploma, he finally got accepted as the second black student to attend Simpson. Learning of his love for plants, one of his teachers suggested that he transfer to the agricultural college of Iowa State University at Ames. As the only black on campus, he faced a considerable struggle to be accepted, but his personal charm and devotion to his studies eventually won him many friends. He excelled in botany and horticulture. His senior thesis, "Plants as Modified by Man," was innovative for the time in its suggestions for using hybridization to select desired features of plants. After graduating, he taught freshman biology at Ames and happened to have as an appreciative student Henry Wallace, who later became secretary of agriculture under Franklin Roosevelt.

African-American educator Booker T. Washington, founder of the Tuskegee Institute in Alabama, asked Carver to become its director of agricultural research. Carver accepted after rejecting better offers elsewhere. He described at the time what became his lifelong guiding principle: "It has always been the one ideal of my life to be of the greatest good to the greatest number of 'my people' possible and to this end I have been preparing myself for these many years; feeling as I do that this line of education is the key to unlock the golden door of freedom to our people" (quoted in McMurry 1981). Carver has sometimes been criticized for being too much a minion of the white community and not active enough in protesting violations of the rights of African Americans. He felt, however, that he could make his best contribution by raising the economic status of blacks through practical education. To impart knowledge of the sciences of agriculture seemed to him the greatest contribution he could make to his people, and he devoted his whole life to this task.

In 1902, he published a pamphlet entitled "The Need of Scientific Agriculture in the South" that outlined the ecological problems of the small farmers and exhibited a faith in progress through science that was at its height at the turn of the century. Carver constructed a makeshift laboratory, making do with very meager resources, to conduct experiments on how to raise agricultural productivity. He developed programs to help small farmers make effective use of their land, persuading them to diversify their crops and to use soil-enriching sweet potatoes and peanuts in place of cotton. Then his attention turned to finding how to extract or produce useful products from these crops, making, for example, vinegar, molasses, and rubber from sweet potatoes. He became known as the "Peanut Man" on account of the more than three hundred products that he synthesized from them—dyes, paste, soaps, and so on. Due to his research, peanuts—once a rarely cultivated commodity—became the South's second-largest cash crop. Carver never attempted to profit personally from his discoveries; although he received offers to join inventor Thomas Edison and automobile maker Henry Ford—with great increase in salary—he turned them down. (He did work with Ford on extracting rubber from goldenrod.) Joseph Stalin even invited him to supervise cotton planting in the Soviet Union, but he stuck to his project of helping the lot of poor blacks in the South. (Carver's simple but sound methods for increasing agricultural productivity would probably have been more effective than Trofim Lysenko's pseudoscientific programs—see 8.2.)

During the 1920s, Carver went to various colleges on lecture tours sponsored by the Commission on Interracial Cooperation and the YMCA and in his quiet way made people more aware of racial injustices. His efforts improved the living conditions of many sharecroppers, and he became a source of inspiration to African Americans. Linda McMurry, in her excellent book *George Washington Carver: Scientist and Symbol,* summarized his greatest contribution as having made "science seem more human and understandable" (vii). Carver's concept of sustainable agriculture and renewable resources was ahead of his time. His advocacy of organic fertilizers and "natural" methods anticipated ideas that only became prominent more than half a century later. One of his principles was that nature wastes not. By applying human intelligence, he believed, natural resources could be used without waste and degradation of the environment. Lectures such as "Does Scientific Agriculture

Pay?" expressed a point of view more commonly associated with contemporary naturalists and ecologists than with agricultural engineers. He had a holistic view of nature. When asked how he could have worked on so many different things, he replied that the projects really were not different, quoting Alfred Lord Tennyson's famous poem "Flower in the Crannied Wall":

I hold you here, root and all, in my hand,
Little flower—but if I could understand
What you are, root and all, and all in all,
I should know what God and man is.

That was, he commented, the kind of truth that the scientist seeks (McMurry 1981, 302). In 1939, Carver was awarded the Roosevelt Medal with the citation: "To a scientist humbly seeking the guidance of God and a liberator to men of the white race as well as the black."

Haber: Chemical Weapons and Gold from the Sea

The story of Fritz Haber (1868–1934) is that of a brilliant German chemist trying to be of service to his country in peace and in war, with very mixed results. Haber rose to scientific eminence in late nineteenth-century Germany, where the efficient organization of scientific education had become a model for the rest of the world. In the early 1900s, he solved the problem of how to synthesize ammonia from nitrogen and hydrogen, by using high temperature and pressure and an iron catalyst. This discovery was of more than academic interest, for agriculture was becoming increasingly dependent on the use of fertilizers, whose essential constituent was "fixed" (chemically combined) nitrogen. This crucial ingredient could be found in natural sources such as bird dung (guano) or nitrates from the Chilean desert, but the existing supplies were rapidly being used up. Nitrogen is present in the air in vast quantities (making up more than three-fourths of the atmosphere), but it is useless in pure form to most organisms. Haber showed how nitrogen could be extracted from the air, and for this work he received the Nobel Prize in 1919. But it was one thing to do this under controlled conditions in the laboratory and another to find a way of doing it on an industrial scale. The engineering problem was solved by Carl Bosch, so the method is now known as the Haber-Bosch process. Haber himself was fond of explaining the superiority that the German chemical industry had attained over its British counterpart. He remarked that the English social system hindered the useful application of science because industrialists aspired to be leisured gentlemen and did not take science seriously, whereas in German industry, scientists were given a leading role (Crowther 1941, 501).

German organization and efficiency were also applied to the funding of scientific research. In 1911, the Kaiser Wilhelm Gesellschaft, a society specifically for the promotion of research, was founded. It was neither a university nor an agency of the state, and it had no responsibility for teaching. The money for its operation was subscribed initially by bankers and industrialists, which enabled it to pursue its own policy independently of the state (despite its imperial title). From its beginning, the society followed a policy of building its "institutes" around an eminent scholar. Two series of these institutes were set up, in theoretical and in applied science. Haber was invited to become director of the Institute for Physical Chemistry and Electrochemistry at Dahlem, near Berlin. He was to remain there, very much the king of his own scientific castle but also a loyal servant of Germany, until nearly the end of his career.

Three years after this confident beginning, World War I broke out, and Haber immediately put himself and his institute at the service of the War Ministry. His previous discovery of how to synthesize ammonia turned out to be the salvation of his country in the first year of the war. Initially, the German high command had thought that its forces could win within months, but by the end of 1914, the opposing armies were bogged down in the trenches, and it became apparent that they were in for a long slog. Nitrates were urgently needed for fertilizers, but the British naval blockade had cut off shipping lines to Chile, where guano was usually collected. If it had not been for the Haber-Bosch process, Germany might well have run out of food. Not that Haber had foreseen all this ten years earlier, but he obviously was well aware of the economic importance of the process; his interest in it was fueled by more than pure curiosity.

Haber's scientific expertise was soon put to direct military use. The German high command consulted him about the possibility of new weapons, and he tried out chlorine gas as an irritant to drive enemy troops out of their trenches. The gas was first used in April 1915 at Ypres, with more success than the local German commanders expected; they were not prepared to exploit the gap created in the opposing front line. But subsequent trials showed the gas to be a clumsy, double-edged weapon, for the prevailing winds tended to blow it back onto German lines. Gas masks for German personnel were thus needed, and Haber was soon working on that problem, too. He was appointed chief of chemical warfare in 1916, having already, to his great pride, been made a captain in the army. He proceeded to develop phosgene and then mustard gas, which tended to disable or kill—which it did to thousands of unfortunate infantrymen. He managed the production with his usual efficiency, and many thousands of tons of poisonous gases were used. Clearly, he put all his expertise and energy into the service of the German military, without any hesitation or scruples. But it seems that his wife disliked the whole idea so much that she committed suicide during this time, while Haber pressed on with his military work.

Haber was profoundly shaken by the German defeat in 1918. The victorious Allies cited him as a war criminal for having invented gas warfare. Coming to his own defense, he replied that air and submarine weapons were equally novel and deadly and that new armaments have always been condemned as barbarous until people get used to them. Germany was faced with crippling financial reparations after the war, and Haber conceived the amazing project of helping his country pay these off by extracting gold from the sea. This was not quite as mad as it sounds, for gold is present in seawater, a huge amount in total, though very dilute. The problem was to find an economical way to carry out this scheme. He devoted much of his ingenuity to this end, but to no avail. He was still a patriot, devoting his scientific expertise to solving his country's problems.

During the 1920s, Haber remained director of his institute at Dahlem, inspiring the work of younger chemists, and had the satisfaction of working in one of the great international centers of science, which at that time included Planck and Einstein as members. He has been described as the most characteristic figure and the strongest personality in German science of that era: "hard, generous, lively," with a driving capacity for organization and resolute pursuit of detail, dignified but with childlike vanity, and courtly manners (Crowther 1941, 496).

Haber's world fell apart around him when Adolf Hitler came to power in 1933 and Jews were systematically expelled from his institute. Although he was part Jewish himself (which apparently was why he had not been promoted further than the rank of captain in the army), he was not dismissed. He felt compelled to resign on principle, however, in protest against the application of a racial criterion for scientific employment. He left his beloved country, and before he died, he found brief refuge in Cambridge, England, the land of his former enemies, ironically enough. It took the rise of the Nazis to make him doubt his previously unquestioned German nationalism. As a scientist who devoted so much of his time and energy to being useful to his country, Haber serves as an example of how one man's patriotism can, quite literally, be another man's poison.

Suggestions for Further Reading

Bradley, D. 1967. *Count Rumford.*
Carter, R. 1965. *Breakthrough: The Saga of Jonas Salk.*
Dickinson, H. W. 1936. *James Watt.*
Dubos, R. J. 1950. *Louis Pasteur.*
Gillispie, C. C. 1970–1980. *Dictionary of Scientific Biography.*
Goran, M. 1967. *The Story of Fritz Haber.*
Hall, A. R. 1962. *The Scientific Revolution, 1500–1800.*
McMurry, L. O. 1981. *George Washington Carver: Scientist and Symbol.*

Science and Money

All scientists are concerned with money, at least to the extent that like other mortals, they have to secure the necessities of life. To acquire scientific qualifications and spend substantial effort on scientific research, one needs time free from shopping, cooking, cleaning, and childcare. So unless a scientist happens to have private wealth, he or she needs an income.

Over and above satisfying their personal needs, scientists have to provide for the cost of their research—which today often involves teams of specialist scientists, technicians, secretaries, computer operators, and high-tech equipment. Since most scientists are attached to universities and the cost of such salaries and hardware may exceed the budget of the average university department, much of the funding must be secured elsewhere. Thus, many scientists now find themselves spending a lot of their time preparing detailed grant applications, many of which will be unsuccessful, given the strength of competition and the limits of available funds. In many fields of science, it is not enough to secure a salaried university position—one must also have excellent "grantsmanship." Without sufficient brilliance, luck, patronage, or influence to succeed in the competition for funding, one may not be able to do much research in present-day science.

In recent times, with the increasing dependence of scientific research on complex and expensive technology, scientists have become more like entrepreneurs. The opportunities for joining private industry and sharing the sometimes substantial profits flowing from applications of basic scientific research has affected the way science is done. Particularly in biotechnology, some scientists have begun to patent discoveries and form companies to profit from them. Money for funding and profit has become an increasingly important influence on scientific research.

7.1 The Funding of Scientific Research

The need to find financial support for science is hardly a new phenomenon. Early scientists also faced problems in securing the requisite funds to pursue their work. Admittedly, some were wealthy enough to indulge their intellectual interests, and a few like Newton were lucky enough to get secure and well-rewarded academic positions (see 1.2, 1.3, and 5.1). But others had to find nonscientific employment and do their research in their spare time. Some scientists supported themselves by finding patronage, selling their expertise to those who had money and power. From those times down to the present day, the social and political commitments involved in obtaining financial support have rebounded on scientists in unexpected ways. We shall look at a variety of cases, from the seventeenth century to the twentieth.

Galileo: Consequences of Financial Need

Galileo Galilei (1564–1642) was one of the founders of modern physics; indeed, he is sometimes described as the first modern scientist, for he developed the mathematical approach to nature that made the big break from preceding Aristotelian ideas and has been characteristic of physics ever since. He laid the basis of terrestrial mechanics, on which Newton built his great synthesis of universal gravitation. Along with Kepler and Newton, whom we considered in 4.1, Galileo used both observational evidence and theoretical argument to support the sun-centered cosmology that Copernicus had proposed in the sixteenth century. For this, notoriously, he was brought before the Inquisition of the Roman Catholic Church, found guilty of heresy, and forced to retract his support of the Copernican system. But this much-discussed episode will not be central to our concern here; instead, we will look at some rather less well known facts about how Galileo managed to support himself financially. This may throw a different light on that famous confrontation between scientist and church.

Galileo's father wished his son to study medicine, but the youth soon developed a strong interest in mathematics and physics to the detriment of his medical studies. His mathematical talent was recognized, but his father refused to let him change subjects because of the poor prospects of earning a living as a mathematician. In 1585, Galileo abandoned his medical studies without having gotten a degree. He supported himself for a few years by giving private instruction in mathematics (and also medical consultations) while using his spare time to write books on hydrostatics and motion. In 1589, he was given a three-year appointment as professor of mathematics at the University of Pisa, but the position came with a very low salary—one-thirtieth of that for a chair in medicine. (Differentials in academic salaries persist, but not quite to that extent!) However, he was ejected from his position at Pisa on the grounds that he flouted conventions of dress and behavior—and probably also because some were jealous of his talent. With the help of influential patrons who appreciated his mathematical genius, he was then able to secure an appointment at the more prestigious University of Padua.

Galileo spent much of his salary supporting his family back in Florence. Seeking ways to supplement his income, he offered private instruction in military architecture, fortification, surveying, and mechanics to the many foreigners in Padua with military interests. In 1597, he devised an instrument that, for sales appeal, he called "the geometric and military compass" for use by the artillery. He employed a craftsman to make these instruments in quantity ("enterprise culture" among scientists is nothing new!). The Venetian senate was glad to use Galileo's expertise for its armaments, and he was happy to pick up the extra cash.

Though amassing a fortune never became for him an end in itself, Galileo did desire a level of income that would secure him enough financial independence to pursue his passion to understand the physical world. He was more worldly than many of his contemporary colleagues and was quite aware of the potential commercial value of some of his discoveries. On hearing of the invention of the telescope in 1608, he worked out its principles for himself and constructed a more powerful instrument, which he presented to the doge (ruler) of Venice. He then demonstrated its magnifying power to some senators assembled at the top of St. Mark's tower and explained how it could be used to discern incoming ships more quickly, an advantage both in trade and war. He was rewarded with an increase in salary and a lifelong tenure for his professorship. After he had used his telescope to discover the moons revolving around Jupiter, Galileo even tried to sell the privilege of naming them to the royalty of Europe. Later, he proposed a method of determining longitude at sea (see 1.2) that depended on telescopic observation of Jupiter's satellites. He tried to peddle this idea to the rulers of Spain and Holland but found no takers; the method was theoretically possible but not practicable for sailors, so for a second time the moons of Jupiter failed to increase Galileo's income!

Despite having achieved a more secure financial position, Galileo was not satisfied with his situation at Padua. Like many scientists since then, he found teaching duties a burden on his time and longed for a position that would free him to get on with his research. He also yearned to return to Florence, where he had spent his boyhood. His chance came when Cosimo de Medici, once one of his pupils and now the duke of Tuscany, offered him an appointment as "philosopher and chief mathematician." Galileo increased his favor with the duke by dedicating to him the popular pamphlet *Siderius Nuncius* (The Starry Messenger), which described some of the former's celestial observations. Galileo's position at Cosimo's court was specially created for him and demanded no routine duties—the duke was rich enough to indulge the whim of having a scientist as part of the palace retinue. All went well for a while, and at the Grand Ducal Court the arguments for and against the new Copernican system of the world were excitedly discussed. But that theory differed crucially from the earth-centered cosmology inherited from Aristotle, which had for centuries been incorporated into Christian doctrine. The church also felt its authority threatened by Galileo's observations of the heavens, both for what was revealed and how. He claimed that the moon had mountains, the sun had spots, Jupiter had moons, and Venus had phases; these facts undercut the traditional distinction between the perfection of the heavens and the imperfect nature of earthly phenomena. And use of the telescope as a source of "revelation" did not require any mediation of church authorities.

Galileo was brought to trial before the Inquisition for defying their ban on assertions of the sun-centered cosmology. He was eventually forced, under threat of torture, to renounce his claim that the earth moves around the sun, and he was put under house arrest for the rest of his life. The tangled history of this confrontation with the power of the Catholic Church has been told and analyzed many times; the interpretation of the episode and some of its historical details remain controversial even now. We shall not try to add to this story here. We note only the economic factor—that Galileo's move from Venice to Florence in his search for patronage brought him nearer Rome and subject to the political power of the papacy and the jurisdiction of the Inquisition. Galileo must have thought that he was gaining for himself the freedom to pursue his research, but by accepting this particular private patronage, he actually exposed himself to political and religious forces beyond his control.

Why Lavoisier Ended Up on the Guillotine

Antoine-Laurent Lavoisier (1743–1794) is known as the father of chemistry for having laid the foundations on which all present-day chemical theory is based. He showed that air and water are not elements (as the Greeks had thought), and he identified many of the basic, true elements such as hydrogen, oxygen, and nitrogen. By demonstrating the role of oxygen in combustion, he effectively disproved the phlogiston theory of combustion widely held in the eighteenth century (see the discussion of Priestley in 4.3). He learned from the experiments of his predecessors, such as Priestley (scientific communication was quite effective even then), but his originality as a theoretician is clear. Lavoisier invented the system of chemical nomenclature for elements and compounds that has long since become standard. But he is also famous for losing his head to the guillotine in the French Revolution. We shall examine here what connection there might be between his scientific achievement and political nemesis.

Lavoisier started with every possible advantage. He was the son of a wealthy Parisian lawyer, and he received an excellent scientific education at the College Mazarin. He started on the professional path by getting himself elected as a member of the prestigious Academie des Sciences at the relatively early age of twenty-five. (The French Academy, unlike the Royal Society of London, had a number of government-funded appointments.) But his ambitions were not satisfied. To secure for himself the freedom that a larger income would give, he made what seemed at the time a shrewd investment by buying himself a position in the Ferme Generale, which was a semiofficial tax-collecting agency. The French kings in those prerevolutionary times (later called the *ancien régime*) delegated the task of collecting taxes on tobacco and salt (as well as customs duties on goods entering Paris) to this private consortium, whose members were able to line their own pockets with a slice of the proceeds. Naturally enough, members of the Ferme were not popular. In 1775, Lavoisier's links with the regime became even closer when he was appointed a commissioner in the French Gunpowder Office and became scientific director of the state arsenal. He actually lived within this institution and had the use of its considerable resources for his chemical research (another early case of military funding for science).

When the Revolution broke out in 1789, Lavoisier supported it at first, having already recognized the urgent need for reform of the old royal system in France. He even played an important role in organizing public finances under the new revolutionary regime. But as the early hopes for peaceful change disintegrated in the face of social chaos, more extreme factions came to power, and the notorious "reign of terror" ensued. At this stage, Lavoisier's strong associations with the institutions of the *ancien régime* made him a natural object of suspicion. He was even accused, absurdly enough, of stopping the circulation of air in Paris, because he had had a wall built as a customs barrier! He was arrested—under the political conditions of the time, his outstanding scientific achievements were no defense. The arresting officer is supposed to have said that "the Republic has no need of experts." Such an attitude of scorn for real science as opposed to ideology appeared again in both Nazi and Communist regimes in the twentieth century. Lavoisier was guillotined in 1794. There could hardly be a more dramatic example of sources of funding rebounding on the scientist.

How Darwin Could Afford to Do His Research

We discussed Darwin's problem with Wallace regarding which of them originated the concept of evolution by natural selection in 5.1. The question here is how he could afford to devote his life to science; the short answer is that he and his wife inherited sufficient private means. His father was a successful country doctor in Shrewsbury, England, who thereby earned a comfortable income. Charles's mother was Susannah Wedgewood, daughter of the famous potter, who was also quite well-off. So the young Charles was able to foresee that his inheritance would provide him enough to live on, and by marrying his cousin Emma Wedgewood he grew even wealthier. He could have been just one more of a familiar type—a Victorian gentleman of private means (supported by a dutiful wife and domestic servants) who could afford to indulge a taste for science in his long hours of leisure (see 1.3). But he became much more than that. How did it come about?

Charles was sent to Edinburgh University to follow the family tradition by studying medicine (his grandfather Erasmus Darwin, also a doctor, had already speculated about evolution.) But he was far from enthusiastic about medicine, being especially put off by witnessing operations performed without anaesthetic. He much preferred prospecting for rocks around the Firth of Forth. But if he was not to follow the family medical tradition, a recognized alternative profession for a young gentleman of his social background was to become an Anglican clergyman, for which a prerequisite was a degree from the University of Cambridge. Darwin embarked on this course but made little of his official curriculum there. However, he persisted in collecting rocks and beetles and studying the biological knowledge of the times, thereby gaining the respect of geologist Adam Sedgwick and natural historian John Henslow.

It was through Henslow that Darwin came by his greatest single piece of luck, namely, the invitation to join the circumglobal survey voyage of the British Admiralty ship H.M.S. *Beagle.* Captain Robert FitzRoy wanted company for the long trip and was prepared to share his cabin with a naturalist—provided that the latter was a gentleman. When the idea was put to Darwin, his first inclination was to jump at the chance; he had already read the account by German explorer and naturalist Alexander von Humboldt (1769–1859) of a journey to South America. But his father thought that such an adventure would be yet another time-wasting avoidance of commitment to a proper career. Accordingly, Charles wrote a letter of refusal. Fortunately for science, the senior Darwin was won over by the arguments of Charles's uncle Josiah Wedgewood. Charles rescinded his refusal just in time to embark on a voyage that was to change his life and the thinking of future generations. Thus, through a series of happy accidents, one of the most important research grants of all time was awarded.

The rest is history, as they say. The energetic young Darwin started out as an amateur scientist without any official qualifications. But he had informally learned much from his distinguished mentors at Cambridge, and there was time on the voyage to make a careful study of Charles Lyell's (1797–1875) newly published *Principles of Geology*, which was sent to him en route. The five-year journey gave him immense amounts of new biological and geological knowledge from the many coastal and island visits and plenty of time to think during the sea passages. He returned as an expert in several fields with new data and new theories to report to the English scientific community.

After marrying his cousin Emma and staying briefly in London, Charles Darwin lived quietly, almost reclusively, in the village of Down in Kent for the rest of his long life. Ill health handicapped him making him periodically unable to function. One diagnosis of his health problems is that he had contracted what is now known as Chagas' disease from the bites of "the black bug of the pampas" while exploring in Argentina. Another suggestion is that Darwin's symptoms of palpitations, fainting, nausea, and fatigue that recurred throughout his life arose from stress caused by personal problems and worry about the reception of his work (Bowlby 1990, 457–466). Darwin conserved his energies, methodically developing his theories and keeping well away from public discussion and controversy.

The crucial idea of evolution by natural selection occurred to him in 1838, suddenly coming to him, he later wrote, as he read a passage in Malthus (see 5.1) on the geometric rate of increase of populations and the subsequent struggle for existence. For many years he did not publish what he realized was a vitally important hypothesis, because he was keenly aware of how much theological, social, and political controversy it would arouse. But eventually he was forced into the open by a letter from Wallace (see 5.1). His *Origin of Species* was published in 1859, and we still live with its reverberations. (Present-day controversies in human sociobiology, for example, are rooted directly in Darwin's theories.) He accepted the obvious implication of his theory—that human beings, like any other living organisms, are a product of evolution, a view that clashed with the traditional biblical account in Genesis. He went on to write a book on the expression of emotion in animals and man. Much of Darwin's work prefigures twentieth-century ecology and ethology (see 9.3).

Here was a scientist blessed with two strokes of good fortune—the chance to join the voyage of the *Beagle* and amass so much new data and the money to spend his time thinking about the implications for years afterward. Many other able scientists have not been so fortunate. As we have seen earlier in this chapter, Galileo had to depend on patronage, and the allegedly heretical nature of his scientific theory put him at the mercy of the powers of his time. If Darwin had lived in a less tolerant society than Victorian England—or indeed, in a less economically privileged position within it—he could well have found himself in similar trouble, for his theory of evolution was at least equally subversive of both common sense and religious thinking in its implications about human nature. Darwin was very fortunate, but it took all of his talents and determination to make so much of the opportunities.

Summerlin: Cheating on the Patchwork Mouse

Almost every scientist nowadays is vividly aware of the problems of getting funding. Money usually has to be obtained from the state, from industry, from research councils, or from private foundations. Persuading such institutions to back one's favored line of research means competing with many other applicants who are trying to do the same thing. Projects must therefore be very carefully presented to emphasize their scientific interest, feasibility, and practical usefulness. The competence of the aspiring researcher is an important factor, too, and the obvious way of demonstrating this is by results already achieved. Even after funding is awarded, its continuation is not open-ended; at some stage, progress will be assessed before further support is given. In such situations, the competition to stay in the game creates strong pressure to come up with positive results—or least promising ones. The immediate motive for work can thus become not curiosity or usefulness but the need to impress the boss of the laboratory and the grant-awarding bodies.

A case in point—an extreme case, but instructive nonetheless—is that of Dr. William T. Summerlin, who was caught engaging in a bit of elementary scientific cheating in 1974. He was a medically trained researcher working at the prestigious Sloan-Kettering Cancer Center in New York under the direction of Robert A. Good. The latter had widely acknowledged scientific superstar status in the highly funded area of cancer research. Directing a large team of assistants, with his name appearing (by the usual convention) on each of their papers, he had become the most-cited scientific author of all time, as measured by quantitative citation indices. He had even achieved the public relations coup of getting his picture on the front cover of *Time* magazine. Good had just moved from Minneapolis to become head of the Sloan-Kettering

Center, bringing with him a large retinue of scientific assistants including Summerlin, whose controversial work he had backed in Minnesota.

Summerlin had been trying to achieve transplants from one animal to another by nurturing the relevant piece of tissue in a special laboratory culture. He had claimed that the tissue would not then encounter the usual rejection by the immune system of the new host. This line of research was thought to be potentially relevant to understanding cancer, for the body's immune system normally succeeds in rejecting and destroying unusual growths, whereas in cancer this defense mechanism breaks down. Any contribution to understanding how the immune system works thus looked promising. However, Summerlin's claimed results went contrary to everything that was believed in immunology, and when other researchers tried to replicate these results, they did not achieve success. A rival in the same lab at Sloan-Kettering was proposing to publish some such negative results and thus throw public doubt on Summerlin's claims and was asking for Good's backing. Good himself was beginning to wonder about the validity of his protégé's work, and relations between the two had deteriorated. Summerlin later said that he was very angry with Good, presumably because he felt that his scientific patron was turning against him.

One crucial morning in March 1974, Summerlin was due to present Good with the results of his latest experiments (in the form of black laboratory mice that had received skin grafts from white mice). As he headed toward his boss's office in the elevator, he used a felt-tipped pen to darken the patches of transplanted skin; such a change in color would suggest that the transplants had "taken." The most charitable explanation for this action is that Summerlin was trying to make the results, which he believed to be positive, look more obviously so to his boss, without intending to fake the essential point. But of course, *any* such unadmitted interference with experimental data goes flatly against the ethics of science. Good did not notice the ink, but a sharp-eyed laboratory assistant did, and having no particular love for Summerlin (personal tensions exist in every lab!), the tale was rapidly brought to the ears of the boss. When Good called Summerlin back in, he admitted the inking and was promptly suspended from work, pending a full investigation.

The subsequent inquiry uncovered a much longer history of sloppy, dubious, and possibly deceptive practices by Summerlin. More serious in scientific terms than the impulsive inking of the mice was his failure to follow standard control procedures when trying to transplant corneas onto the eyes of rabbits. Apparently, Summerlin had even presented certain cases as successful transplantations when no operation had been performed at all. As a result of the inquiry, he was dismissed from the Sloan-Kettering Cancer Center. Rather insultingly, he was described as "emotionally disturbed" and "not fully responsible for his actions"—as if the authorities were unwilling to accept that deliberate faking had been going on in their labs. He was given a year's leave on full salary, and he later returned to medical work in a rural area.

Good was also implicated because he had gone out of his way to recruit Summerlin on the strength of his surprising claims about transplantation and had encouraged his work even when others reported failure to replicate the claimed results. Good was found innocent of any misrepresentation but guilty of overhastiness in publicizing claims he had not checked. Like others in positions of administrative power in science, he had been too busy to oversee properly the work of those for whom he was responsible and to whose papers he put his name. And it seems there were those within Sloan-Kettering who were not unhappy to see their imperious, go-getting president brought down a peg or two. The prestigious cancer center itself was also adversely affected when the story of the falsification was leaked to the media.

The more perceptive scientific journalists drew wider conclusions about the way in which science was being directed, funded, and presented to the public (at least in the United States). James Watson's diagnosis was that too much money and effort was being put into applied research aimed directly to find the cause of cancer, to the detriment of basic research in biology and biochemistry, which, he argued, is the only proper basis for full understanding and reliable cures. In the eager search for practical results, publicity was being given to sensational but unconfirmed claims. In the competition to gain and retain funding, junior scientists like Summerlin were under undue pressure to produce quick results and could thus be tempted to fake their work.

7.2 Profiting from Science

Scientists since Galileo's time have taken opportunities to make money from applications of their work. In modern capitalist economies, it is quite common for research that has an intrinsic intellectual fascination also to promise considerable personal or institutional profits to those who can make a breakthrough and patent it first. The profit motive—in some cases, the lure of immense riches—can change the normal academic way of conducting scientific research, as we shall now see.

The Curies' Refusal to Make Profits

According to what now seems an old-fashioned tradition, it was often thought unethical for scientists to profit personally from their discoveries. A very clear example of high-minded refusal to make money from scientific research was set by Marie and Pierre Curie. Soon after their identification and separation of the previously unknown radioactive element radium (see 4.2), it was realized that the new substance could have medical uses in the treatment of tumors and certain forms of cancer, and so there was an immediate demand to produce it in larger quantities. At this stage, the Curies could have ensured for themselves a considerable personal fortune if they had chosen to patent their painstakingly devised technique for purifying radium. But in one of the most extraordinarily selfless decisions of all science, they did not do so, despite the fact that they were not at all well off, had just started a family, and still lacked proper equipment to pursue further research. They felt very strongly that radium belonged to science, not to themselves personally. According to Eve Curie, her mother said: "Physicists have always published their researches completely. If our discovery has a commercial future, that is an accident by which we must not profit. And radium is going to be of use in treating disease. . . . It seems to me impossible to take advantage of that" (Curie 1937, 204).

Even the very valuable gram of radium that they labored so hard to produce at their own expense—and that was legally their own property—was given to the cause of scientific research.

How many scientists would follow the Curies' example nowadays? Many would find them irrational in refusing to profit from their own work. But what should be the ethical and legal constraints on scientists? The question is not an easy one; perhaps we can throw some light on it by considering further case studies.

How Nobel Got the Money for the Prizes

A sharp contrast can be seen between the Curies' attitude and that of Alfred Nobel (1833–1896). Because of inventions such as dynamite, Nobel amassed an immense fortune, which provided the endowment for the Nobel Prizes. For better or for worse, these awards have become a standard measure of scientific preeminence, both for the public and for scientists themselves (see 5.1). According to our definition given in 3.1, Nobel hardly counts as a scientist, for he did not contribute to scientific theory or make much attempt to do so. But through an unorthodox education, he attained considerable competence in chemistry, and for much of his life he used this knowledge and skill to practical effect as well as to his own profit.

Nobel's father and brothers were inventors and arms manufacturers, and Alfred seems to have easily accepted the role of armament producer without national allegiance, prepared to sell his expertise to the highest bidder. In addition to his precocious technical ability, he was keenly aware of the value of patenting new inventions as quickly as possible, and 355 patents were awarded to him in his lifetime. He got involved in several priority disputes, the first of them with his own father!

His first major invention—the Nobel patent detonator—arrived on the market in 1863. He then set out to manufacture nitroglycerine in the liquid form in which it was already known, and he founded the first limited company in explosives. But the stuff was highly unstable, and in 1864 the factory blew up, killing his younger brother, Emil, and four others. Undeterred by this disaster or by the Swedish government's ban on manufacture and storage of nitroglycerine in a populated area, he proceeded with experiments on a barge. By chance, Nobel happened upon a combination of nitroglycerine oil and kieselguhr, a kind of fossil silica, that yielded an explosive with a convenient plastic form. He called this new substance "dynamite," and it turned out to be his greatest invention.

Nobel's explosives company grew swiftly. Patents were granted on dynamite in Britain in 1867 and in the United States the following year, and soon the profits were flowing in. Nobel's further experiments produced a more powerful version of dynamite called blasting gelatin and, later, ballistite, a smokeless powder specially designed for wartime use. Factories to make the explosives were established all over Europe; their products were used by all combatants in subsequent wars as well as for more constructive purposes like blasting tunnels through the Alps. Nobel rapidly became a millionaire. He devoted himself assiduously to the extension of his business, even after he was already very rich. Making money seems to have been one of his strongest motivations, but the other clearly was his technical interest in new inventions. He continued experiments on silent guns, shell fuses, propulsion charges, rocket projectiles, and so on even toward the end of his life. Not all his work was of a military nature, however. Later he experimented in optics, electrochemistry, biology, and physiology, and this work contributed to the subsequent development of artificial rubber, leather, silk, and semiprecious stones, which are made from fused alumina.

Three years before his death, having no children, Nobel expressed a wish that his immense fortune should be used to set up a prize for the person who had taken the greatest step toward the ideal of peace in Europe. Nobel's will actually established five prizes to be awarded annually: four "for the most outstanding work of an idealistic tendency" in physics, chemistry, physiology or medicine, and literature and one, the Peace Prize, "for the person who shall have done the most or the best work to promote fraternity between nations for the abolishment or reduction of standing armies and the holding and promotion of peace congresses." A prize in economics has since been added to the list. The prizes have long been regarded as the most coveted (as well as the best-endowed) honors in science.

Alfred Nobel was a researcher whose prime motivation seems to have been to make money, though he was also fascinated with the technical details of his craft. For many in his time (and ours), it was an accepted practice to patent and sell new scientifically based inventions, making money from them regardless of the uses in peace or war to which they might be put.

Chain's Defense of the Drug Industry

The industrial affiliations of Sir Ernst Chain (1906–1979) serve to underscore the mixed motivation of scientists who get involved with commercial matters. His father was a chemical engineer in a major German company, so he was presumably attuned to the industrial application of science from the start. Like many German-born Jewish intellectuals, Chain sought refuge abroad from Nazi persecution. After studying biochemistry at Cambridge, he was recruited in 1936 into Howard Florey's research team at the University of Oxford's School of Pathology. When small grants from the British government ran out, Florey approached the Rockefeller Foundation for more substantial funding, which supported the famous research that resulted in the mass production of penicillin (discussed in 4.3 and 5.3).

A financially significant episode in this story occurred when Chain pointed out the necessity of patenting the production process that the Oxford team had developed. He realized that if he and his colleagues did not do so, someone else surely would, and he argued his case vigorously with high officials of the Medical Research Council and the Royal Society. But in the England of the time (unlike the Germany from which Chain had come), a sharp division existed between pure and applied science and between academia and industry. Any suggestion that university scientists might take a commercial attitude toward their work was anathema, and Chain was told so in no uncertain terms (how things have changed!). As a refugee with uncertain status in Britain, he had to swallow his pride (he wrote afterward that he encountered many bitter fights on the patent issue), but his judgment was vindicated when the British team eventually had to pay royalties to American companies in order to use methods they themselves had first developed.

After the war, Chain again found himself frustrated by the lack of funds for the research he wanted to do. In 1948, he accepted an invitation to set up a new Department of Biochemistry at the Italian State Institute of Health in Rome. The decisive factor behind his move seems to have been the promise of facilities and financial support that Britain could not offer. Once again, Chain was very much alive to the commercial possibilities of his research. For some years after the production of penicillin, it was possible to find other antibiotics by systematically searching for fungi and bacteria that might produce them. But as such sources began to be exhausted, the pharmaceutical companies began wondering how to develop new drugs. At this stage, the Beechams Group in Britain sought advice from Chain, and he became a consultant for the company. He recommended trying to make semisynthetic penicillins, that is, chemically modified forms of the original organic compound with improved therapeutic power. Two employees of the Beechams Group came to Rome to work in Chain's laboratory, and as a result, the British firm was able to patent several new antibiotics that proved more effective than those sold by the larger companies.

In 1964, Chain returned to England to reestablish the Biochemistry Department at Imperial College in London. The cost of the equipment he proposed was unprecedented in university research (except in nuclear physics), but he was able to use his now extensive influence to conjure large sums from sources such as the Wolfson Foundation. This gave his department a financial independence that enabled him to be very much master in his own house; staff who did not like his style had to adjust to it or find other employment. In this position of power, he was emboldened to express scientifically maverick views in later life, doubting the usefulness to biology of the swiftly developing research into the structure of DNA and other biological molecules. He even cast aspersions on the theory of evolution itself, perhaps because of his increasing fondness for the Jewish religion of his youth.

Chain's links with the pharmaceutical industry were a cause of difference with those of his colleagues who continued to believe in the "purity" of university science. He regularly worked as an external consultant, and his department in London collaborated closely with industry, a policy he defended vigorously in public lectures in 1963 and 1970. And of course, penicillin was a huge medical success story. Although worries have since arisen about overprescription of antibiotic drugs and the resultant development of resistant strains of bacteria, there can be no doubt that these drugs have greatly reduced human suffering. In this case, the commercial incentives that motivate the pharmaceutical companies harmonized fairly well with the medical aim to promote human health as well as with the scientific curiosity of biochemists.

In another case, however, we find these motives tearing apart, with tragic consequences. In the early 1960s, a new drug called thalidomide was vigorously marketed in many countries as a harmless sedative or sleeping pill. But it was quite soon noticed that women who had taken it during pregnancy were giving birth to grossly deformed infants, with limbs so undeveloped in some cases that hands and feet appeared to grow directly out of their bodies. The causal connection seemed obvious from the start, for such abnormalities are very rare yet increased dramatically in just those countries where thalidomide had been sold—and within a year of its introduction. Interviews with the unfortunate mothers and their doctors or pharmacists soon revealed that they had taken the new sedative in the early stages of pregnancy.

In most countries, thalidomide was quickly withdrawn from circulation, though not without complaint from the drug companies, which were reluctant to give up a promising source of profits and were extremely reluctant to admit any causal connection or associated legal responsibility. (In the United States, in one shining example of proper caution by the Food and Drug Administration, approval of thalidomide had been blocked because of suspicions about its safety.) The most determined and lengthy resistance came from the West German company Chemie Grunenthal, which had developed thalidomide in the postwar period, when there had been remarkably few legal controls on the marketing of drugs in Germany. When the toxicity of thalidomide became obvious to all except the companies, anguished parents began to demand compensation, and the matter came before the courts in many countries.

This is where Ernst Chain appears in the story. The most prolonged legal case, extending over several years and ending rather unsatisfactorily with an out-of-court settlement, took place in West Germany, where Chemie Grunenthal fought every inch of the way, using its considerable resources to hire clever lawyers and expert witnesses from the scientific community. One of these was Chain. At the trial, he declared melodramatically that the existence of pharmaceutical science itself was threatened by the atmosphere of suspicion that the thalidomide controversy had caused—a remark that hardly suggests neutrality in the case. On the crucial issue of causality, Chain claimed that the clinical trials conducted before marketing thalidomide had proved its harmlessness. This was actually an even stronger line than that taken by the defense, which claimed only that there had been no definite proof of toxicity. Under cross-examination, he was forced to say that a causal relationship, in the strict sense on which he was basing his arguments, did not exist even in other cases where the medical profession generally recognized one (for instance, between diabetes and polyneuritis). He would admit only that there were "indications" of a connection in such cases. Confronted with a quotation from his lecture of 1963 in which he said that the malformed births *were* caused by thalidomide, Chain could only say that the lecture was never intended to be used as evidence, and (incredibly for such a strong-willed man), he suggested that he had been influenced by the press. He went on, in what seems to have been mere bluster, to aver that from what he knew of the pharmaceutical industry, investigations were carried out with the utmost precision. He also claimed that he had known Dr. Muckter (the scientific director of Chemie Grunenthal, who had resisted advice to withdraw thalidomide from the market) for ten years.

Wisely, the court decided that the standard of proof of toxicity demanded by Chain, which would amount to mathematical certainty beyond any possibility of falsification, was far too high for legal purposes (and they could have added, for medical and scientific practice also). The distinguished Professor Chain was surely involved here in a bit of special pleading for his friends in the German pharmaceutical industry. It is sad to find a scientist who had achieved so much committing himself at the end of his career to defending the indefensible. The moral surely is that since profitmaking easily becomes a distorting motive in science, we should be wary that it is kept within the bounds of ethics and well as legality.

Gilbert's Commercialization of Genetic Engineering

The career of Walter Gilbert as scientist and entrepreneur exemplifies in a dramatic way the recent commercialization of molecular biology. In the 1960s, he applied his considerable talents to this swiftly growing field, with such success that he was appointed American Cancer Society Professor of that subject at Harvard and won the Nobel Prize for chemistry. In the mid-1970s, gene-splicing techniques were developed that offered practical possibilities of genetic engineering—deliberately changing genes. The DNA within bacteria, for example, might be modified to produce useful biological substances. Some of these substances might occur naturally but only in limited quantities (such as interferon or human growth hormone), whereas others might be new forms more suitable than the naturally occurring ones for medical, industrial, or agricultural purposes. The distinction between academic research and applied work was becoming unclear, and it was obviously tempting for scientists to make money from their discoveries. For a long time, scientists had undertaken part-time consultancy for industrial companies or governmental agencies, earning fees as a useful supplement to their university salaries. But now many of those working at the cutting edge of molecular biology were founding small genetic engineering companies with the help of venture capitalists. Walter Gilbert was one of the first to set this trend, being involved in the formation of Biogen in 1978 and assuming positions as its acting president, then chairman of its scientific board, and finally cochairman of the board of directors. The Biogen company supported Gilbert's research at Harvard on the bacterial production of insulin.

As funding for research became increasingly hard to find, universities were tempted to try to make a direct profit from the work of their own scientists. By 1977, Harvard University had an office to apply for patents arising from the work of faculty members. Three years later, the Harvard administration proposed setting up a company to develop the work of one of Gilbert's rivals. The proposal was received with a storm of protest by faculty members as well as from the press and the wider community. The general opinion was that this move could be the start of a process by which universities would become almost indistinguishable from commercial corporations, and some observers worried that the university might give special treatment to those of its staff who were involved in companies in which it had a financial interest. So the idea was not pursued in that form. Instead, Gilbert's example has been widely followed. Scientists formed their own companies, either in some loose connection with their institutions or entirely privately. But there are obvious problems with combining the role of an academic scientist hired to do teaching and research with that of a director aiming at financial success for a private corporation. Conflicts of interest and time are inevitable. For a while, Gilbert tried to have his cake and eat it, too, working as an office-bearer of Biogen while remaining a university professor. But the Harvard authorities could not for long tolerate him running his private company from his university office. So he resigned his chair and left the academic world to become a full-time industrialist.

Gilbert's name hit the headlines again in 1987 when he made a breathtakingly audacious proposal to undertake the sequencing of the human genome as a privately financed project. To this end, he established a new company, the Genome Corporation, and his plan was that the company would copyright each section of human DNA as it was laboriously decoded in the company's laboratories. Any subsequent user of the information thus discovered would have to pay a royalty to Gilbert's corporation. But would the courts recognize the validity of a patent or copyright that would apply not to a technical device or process but to the information naturally contained within human genes? Some American lawyers thought so (patents on genetically engineered organisms had been allowed already). But even if this were legally possible, would it be morally acceptable for a private company to make a profit from scientific knowledge? Suppose the Genome Corporation came up with a cure for cancer—would it then demand a fee every time the relevant genetic information was used? Many people argued that scientific knowledge belongs to humanity as a whole, not to a particular group of entrepreneurs. And it was pointed out that the basic biochemical theory and technique on which the Human Genome Project was based had been developed with funding from many public bodies and medical charities. So why should one corporation be allowed to step in at a late stage and make a killing from taking the work one step further?

The Human Genome Project is now being pursued with public funds organized by U.S. government agencies such as the National Institutes of Health (see 9.2). But the fact that Gilbert's proposal was made at all—and taken seriously for a while—shows just how far the commercialization of science has gone.

Fleischmann and Pons: Cold Fusion and Patent Lawyers

On March 23, 1989, the University of Utah held a press conference at which the announcement was made that one of its professors, Stanley Pons, working in collaboration with Martin Fleischmann, formerly of the University of Southampton in England, had induced nuclear fusion at room temperature. What they had done was to pass an electric current through palladium electrodes immersed in deuterium, or heavy water (that is, water in which the hydrogen atoms had an extra neutron in their nuclei). The claim was that as the deuterium atoms crowded inside the palladium metal under high pressure, they fused into helium, releasing both neutrons and heat energy. If this were true, the implications would be staggering, both for physical theory and for energy supply. For decades since the theory of nuclear fusion had been developed and then spectacularly applied in the hydrogen bomb, huge scientific resources have been devoted to seeking a new source of energy by inducing fusion in a controlled way. This research has meant trying to replicate on earth the nuclear processes that are known to power the sun, at fantastically high temperatures and pressures. But so far the energy input in these costly experiments has been greater than the output. Existing physical theory rules out fusion at ordinary temperatures—but if it were possible after all, perhaps the world's energy problems could be solved cheaply and cleanly.

The normal way of announcing a scientific discovery is to publish a technical paper in one of the established scientific journals. The reputation of such periodicals depends on the system of peer review, by which papers are submitted to the judgment of experts in the relevant area and may well be revised in the light of their comments. Announcements to the media are not usually made until the scientific paper has been published, and many science journals have a policy of not accepting anything that has been previously published in a non-peer-reviewed form. But in this case, Pons and Fleischmann made their claim to the media first, without giving sufficient details of their experimental procedures for other scientists to attempt proper replication. What they presented as proof of fusion taking place—the amount of heat generated—was at best only indirect evidence. To demonstrate fusion conclusively, it would have been necessary to detect the emission of neutrons, which they had not done. Fusion is a physical process, not a chemical one; Pons and Fleischmann were not physicists, but they were professors of chemistry and electrochemistry, respectively, with established reputations in their own fields. Though they were not originally household names (only becoming well known because of the controversy aroused by their press conference), their scientific reputations gave their claim some credibility, despite the unusual context of its announcement.

The cold fusion claim caused a considerable stir in the scientific community. And—unusually for a scientific report—it was noticed in the financial world, with both the *Wall Street Journal* and the *London Financial Times* quickly rushing to publish articles about it. The price of palladium shot up. The U.S. Congress and even the president himself took an interest, and urgent scientific studies of cold fusion were commissioned. The military was intrigued by the possibility of a ready supply of tritium (another isotope of hydrogen and an expected by-product of cold fusion), because of its use in keeping nuclear warheads operational.

Over the next few months, scientists of all grades, from undergraduate to professor, in many institutions around the world feverishly strove to replicate the claimed results. The situation was confused for quite some time. A few claimed success, some reported failure, and others changed their minds from one day to the next as a variety of explanations were suggested for both negative and positive results. Much of the trouble arose because Pons and Fleischmann had not supplied sufficient details of what they had done, so experimenters did not know exactly what conditions they were supposed to be setting up (some of them were reduced to scanning videotapes of the press conference to try to pick up clues!). A straight answer was never given as to whether the obvious control experiment, using ordinary water instead of the heavy variety, had been performed. There were difficulties in detecting whether fusion was really taking place. Could the "excess" amounts of heat be accurately measured, and could they be explained in other ways? Conclusive proof of fusion rested on the detection of small quantities of fusion products such as neutrons and tritium, but how could "background" sources of neutrons or contaminations of tritium already in the apparatus be ruled out?

As the dust cleared over the next year or so, a scientific consensus emerged that no clear evidence existed that fusion could occur in the kind of experimental setup that Pons and Fleischmann were using. They surely spoke in good faith—there is no reason to doubt they *believed* that they had achieved cold fusion or at least that they had sufficiently good evidence to go public with the claim. The question that arises is why they bypassed the normal peer review procedure involved in scientific publication and went to the press with a somewhat vaguely specified and unconfirmed claim, thereby putting their own scientific reputations at risk and wasting the time of many scientists who made attempts at replication.

It seems the answer lies in the very high stakes in fame and profit that would have been up for grabs if cold fusion had been verified. A Nobel Prize would surely have been awarded for the discovery. And if a patent could have been taken out on a new energy source for the whole world, the resulting profits would have made even Nobel's fortune look like peanuts. Thus, to the usual competition for priority in scientific discovery was added the lure of immense riches. Behind the scenes, a proposal had been made to divide the prospective royalties three ways: to the University of Utah, to its Department of Chemistry, and to Pons and Fleischmann themselves (even a sixth of the profits would presumably have made them multimillionaires). Naturally, many who thought they were in the running for a patent were afraid that others might get one first, which greatly intensified the usual race for priority in scientific discovery. The University of Utah engaged in some acrimonious competition with Brigham Young University (coincidentally, also in Utah), where Steven Jones had already been experimenting on related topics and was about to publish. Pons and Jones had accused each other of stealing ideas, and at a meeting in the presence of their respective university presidents (following the example of Darwin and Wallace—see 5.1), they agreed to submit their papers for publication simultaneously on March 24, 1989. Yet it seems that the temptations were just too great for the administration of the University of Utah to keep to that agreement, and Pons and Fleischmann were pressured into the notorious press conference on March 23.

For a while, hopes in Utah ran high, and there was heady talk of their university becoming the richest in the world. In a hearing before a congressional science committee, a business consultant hired by the university tried the ploy of saying that unless the government gave the university $25 million to get started on commercial cold fusion reactors (without waiting for confirmation by the scientific establishment), then the Japanese would get in on the act first. Congress was either too shrewd, or too slow, to fall for that one. But the state government of Utah decided to spend $5 million to start a Cold Fusion Center. Upon hearing the objection that the scientific journal *Nature* had not yet accepted the paper submitted by Pons and Fleischmann, the governor's chief of staff, Bud Scruggs, growled, "We are not going to let some English magazine decide how state money is spent!" (quoted in Close 1990, 12). But as 1989 passed by, fewer and fewer investigators claimed they had been able to make cold fusion work; it seems that nature, rather than *Nature*, gave the comeuppance to greed.

The scientific question of cold fusion cannot be regarded as closed (*no* scientific issue can ever be regarded as completely and utterly closed—that is the eternally provisional nature of scientific theory; see 1.1). Pons and Fleischmann have since secured funding from Japanese companies and are pursuing their research with less publicity in the south of France. Some experimenters report having replicated the mysterious excess heat held to be evidence of cold fusion, and venture capitalists are risking their money on further work. Pons and Fleischmann have been forced by criticism to retract any assertions about neutron emission. They have stuck to their basic claim that cold fusion happens, which they defend by postulating unspecified changes in accepted physical theory. Some other scientists speculate about hitherto unknown nuclear processes that may be involved. We must wait and see—science and its technological spin-offs are inherently unpredictable.

It has been remarked that the belief of the few remaining adherents of cold fusion has become like a religious faith, unaffected by the usual processes of scientific experiment and argument. But it is not obvious that the "faith," if such it is, is all on one side. The adherents of the orthodox scientific view that nuclear processes are impossible at room temperature are not totally disinterested (as in many other scientific controversies at the cutting edge of research). They have invested their professional credibility in supporting a particular view, and they do not wish to see it upset. In this case, there are also financial interests on the side of "orthodoxy," for billions have already been spent on hot fusion research, and many people's jobs depend on this. Allegations (difficult to prove or disprove) have even been made that data have been fudged in an attempt to debunk the claim of cold fusion.

It is clear, however, that the original announcement made in 1989 was premature and lacked sufficient scientific backing. And determining how much the two scientists were motivated by the lure of riches or fame or whether their better judgment was overcome by pressures from university administrators, patent lawyers, and the politicians of Utah may be a difficult task. In this case, it does seem as though the financial interests of institutions diverted scientific practice from its normal pattern.

Recrimination finally broke out in Utah. The Cold Fusion Center closed down two years after its confident inception, and the university president came under attack for spending millions of dollars on an abortive project. The Faculty of Science voted to ask the state's governing board for higher education to examine his competence, and he decided to take early retirement. At one stage, Pons even threatened his physicist colleague Michael Salamon with legal action over a critical article the latter published in *Nature*. Pons's attorney demanded that Salamon retract the paper because of its inaccuracy, threatening that any damages suffered by his client would not be tolerated. So not even scientific papers are to be free from American litigiousness! Later this threat was retracted, but that it was issued at all is indicative of the competitiveness and commercialization of present-day science. What price pure scientific curiosity now?

Suggestions for Further Reading

Clark, R. W. 1985. *The Life of Ernst Chain: Penicillin and Beyond.*
Close, F. 1990. *Too Hot to Handle: The Race for Cold Fusion.*
Desmond, A., and J. Moore. 1991. *Darwin: The Life of a Tormented Evolutionist.*
Drake, S. 1978. *Galileo at Work: His Scientific Biography.*
Evlanoff, M., and M. Fluor. 1969. *Alfred Nobel: The Loneliest Millionaire.*
Hall, S. S. 1987. *Invisible Frontiers: The Race to Synthesize a Human Gene.*
Huizinga, J. R. 1993. *Cold Fusion: The Scientific Fiasco of the Century.*
McKie, D. 1952. *Antoine Lavoisier: Scientist, Economist, Social Reformer.*

8 Scientists and the Totalitarian State

Scientists who find themselves living under a dictatorship or any other sort of totalitarian regime face some hard choices, especially if the nature of their work is relevant to the interests of the state. They may serve the government loyally; they may even exploit the political situation for their own advancement. Some may choose to conform outwardly while not committing themselves to the regime insofar as they can avoid doing so, whereas others may to some extent evade or subvert its purposes in clandestine ways. Occasionally a scientist comes out in flat opposition to his or her government's policies, expressing dissent at great personal cost. Examples of each kind of behavior follow.

8.1 Scientists in Nazi Germany

Planck: The Scientist Who Stayed at His Post

When Hitler took power in 1933, German scientists were faced with a stark dilemma: to continue their work under the new regime (and probably be induced to serve its ends) or to leave the country and become refugees. As we have seen in 6.2, some, like Fritz Haber, elected to leave after a lifetime of patriotism. But other scientists, such as the distinguished physicist Max Planck (1858–1947), chose to stay at their positions and try to protect German science from the excesses of Nazi rule. Planck was an old-fashioned, pre-Nazi nationalist like Haber, strongly imbued with the ideal of loyalty to the fatherland and the Kaiser. Yet everyone who knew him has attested to his honest, modest, and gentle character; he has even been described as something of a saint. Like Einstein, he approached science in a quasi-religious spirit, seeing the laws of nature as the expression of a rational world order, although he did not believe in the personal God of Christianity. Einstein's striking words about the few truly devoted worshipers in the temple of science (quoted in 3.2) were written with Planck in mind.

In 1900, physics had entered a new era of theoretical developments, beginning with Planck's theory of the quantum—the idea that energy is not a continuous quantity but comes in discrete, indivisible packets. Planck developed this radically new idea to explain the failure of the classical radiation laws expressing the radiation of energy from a black surface (an ideal black body radiator) as a function of temperature. The equations predicted that the energy radiated would increase without bound as the wavelength of light decreases toward zero. To avoid this "ultraviolet catastrophe," Planck derived a law that fit the experimental data by introducing a radical quantization of energy exchange that marked the beginning of quantum physics. He also insisted (contrary to antirealist or instrumentalist interpretations of scientific theories) on the real existence of tiny, invisible atoms in order to explain the observational evidence. Planck's new physical ideas were taken up by Einstein in one of his famous papers of 1905, and for the rest of his life, Planck was at the forefront of debate in physics, much of it with Einstein himself. It was Planck who persuaded Einstein to join the Physics Institute of the Kaiser Wilhelm Gesellschaft, despite Einstein's reluctance to live in Germany, by arranging for him a special research professorship free of normal duties.

Planck was caught up in the chauvinism that accompanied the outbreak of World War I and enthusiastically supported his country's cause throughout. Afterward, he helped to sustain German science through the economic and political difficulties of the Weimar period. He induced Einstein to stay in Berlin through the 1920s even though rising anti-Jewish sentiment was engendering hostility to Einstein's scientific theories. At one point, Planck arranged a public discussion between Einstein and his chief opponent in German physics (Philipp Lenard, whom we will consider next), but that did not stop the anti-Semitic campaign in science, which had deep roots in German society. (Even before 1900, Planck had opposed official anti-Jewish discrimination on at least two occasions.)

When Hitler became chancellor in 1933, Planck was secretary of the Prussian Academy of Science and president of the Kaiser Wilhelm Gesellschaft. It seems that whatever his reservations about the Nazis, his patriotism was sufficiently unshaken for him to feel that it was his duty to remain at his post and do his best for German science in the new political situation. When Einstein, by then a refugee in the United States, publicly criticized the Nazi government for abandoning civil liberty, tolerance, and equality before the law, Planck felt alienated from his former colleague as well as from the Nazis. He wrote to Einstein: "There has taken place a collision between two conflicting views of the world. I have no understanding for either. I feel remote from yours, as you will remember from our talks about your propaganda for refusing military service." Einstein, for his part, never forgave Planck for not publicly standing up for his Jewish colleagues. Planck did try to act behind the scenes, however. He pleaded personally with Hitler for the reinstatement of Fritz Haber, praising him and other Jewish scientists, but the Führer only flew into a rage; Planck had to take his leave, convinced of the impossibility of rationally influencing Hitler. He showed enough courage to go to a memorial ceremony for Haber two years later, although the government had banned public employees from attending it. Apart from this, he refrained from public protests but did his best to prevent further dismissals of Jewish scientists.

Planck's only surviving son was arrested for conspiracy in the plot against Hitler's life in 1944 and was executed in the closing weeks of the war. Planck wrote: "My sorrow cannot be put into words. I struggle for the strength to give meaning to my future life by conscientious work" (Heilbron 1986, 195). Perhaps some meaning was supplied when the victorious Allies allowed the Kaiser Wilhelm Gesellschaft to be refounded in 1946 as the Max Planck Gesellschaft, named in honor of this outstanding physicist who compromised with the Nazis while trying to preserve German science from their interference.

Lenard and Stark: The Nazi Movement for "German Physics"

Philipp Lenard (1862–1947) and Johannes Stark (1874–1957) were two German experimental physicists of sufficient achievement to win Nobel Prizes for their early work. In the 1920s, after defeat in World War I and the financial reparations exacted by the victors, extreme nationalism and anti-Semitic feeling began to run high in Germany, affecting many scientists, too. Lenard and Stark became the leaders of a group of physicists who called themselves "national researchers" and led an extraordinary movement to develop a specifically German, or "Aryan," physics. Einstein's theory of relativity was dismissed as "Jewish world-bluff," and German physicists who accepted the new concepts of relativity and quantum mechanics were subjected to threatening criticism. Nationalistic, even racial, criteria were applied even within the very content of scientific theory. It is instructive to examine how such a dramatic perversion of science could have happened, however brief its influence. But our discussion of Burt in 5.2 suggests that it is not only in totalitarian societies that we can find social, political, or ideological factors entering into supposedly objective, scientific theorizing.

In the 1890s, Lenard had worked under Heinrich Hertz, the discoverer of electromagnetic radiation, and he took charge of publishing his distinguished mentor's works after the latter's early death. But three decades later Lenard was in the grip of racist ideology and claimed to find taints of "Jewishness" in Hertz's theoretical works. Lenard's early work was on cathode rays; he established that they consist of negatively charged particles. He aggressively disputed priority of discovery of these "electrons," as they were called, with Sir Joseph John Thomson (1856–1940) of Cambridge University in England, who is standardly credited with their discovery. In 1905, Lenard was awarded the Nobel Prize for his cathode ray experiments and was once lauded as "the most celebrated scientist in physical research." Einstein, however, did not concur in this opinion, describing Lenard's lecture on "the abstruse ether" as almost infantile.

Lenard was involved in the general wave of blind nationalism that accompanied World War I. But he took it to greater extremes than most, alleging that British scientists had systematically plagiarized the work of German colleagues. After the war, he incited his students to work against the government of the new Weimar Republic, which he saw as dishonoring Germany by not resisting disarmament. When he refused to close his laboratory for the funeral of a murdered Jew in 1922, a crowd of angry Social Democratic supporters stormed the building, and he had to be taken into protective custody. This literal entry of politics into the scientific laboratory was an ominous sign of times to come, when the Nazi side would do the invading.

Lenard volunteered to head the movement against Einsteinian physics, and he took part in an acrimonious public debate in 1920, at which Max Planck as chairman just managed to prevent an uproar. Besides anti-Semitism, Lenard's attacks on Einstein may have been motivated by his inability to follow the difficult mathematics involved in the new physical theories. His talents were experimental, and he distrusted abstract theorizing. In the context of Germany during the 1920s and 1930s, Lenard's nationalism took a racist form, which showed itself most blatantly in the title of his book *Deutsche Physik*, published in 1937. In the preface he wrote:

> "German physics?" one will ask—I could also have said Aryan physics or physics of the Nordic man, physics of the reality explorers, of the truth seekers, the physics of those who have founded natural science. "Science is and remains international!" someone will reply to me. He, however, is in error. In reality science, like everything man produces, is racially determined, determined by blood. (Beyerchen 1977, 125)

Hitler was delighted to find such a devoted supporter among the scientists, and Lenard became his "authority" in physics. But in fact, German science had been severely weakened by the expulsion of so many of its ablest scientists, and the aging ideologue (Lenard was already seventy when Hitler came to power) proved to be of no practical use to the Nazi war machine.

Johannes Stark's career shows many similarities to Lenard's. He also was a talented experimental physicist but was no whiz at higher mathematics. He discovered a Doppler effect in "canal rays," detecting it in the hydrogen spectral lines, and although he first tried to use this as evidence in favor of Einstein's theory of special relativity and Planck's quantum hypothesis, he later turned strongly against both theories. Some of his early physical hypotheses were shown to be wrong by German physicist Arnold Sommerfeld (1868–1951), and in a pattern that was to recur, Stark reacted with enmity. He seems throughout his life to have been an impossible character to work with. His superior at Hannover had to appeal to the Ministry of Education to transfer him elsewhere, and he got into bitter dispute with the faculty at Goettingen when applying for a chair there. He quarreled with his colleagues at Wuerzburg so intensely that he ended up resigning his professorship. He could afford to move from university to university because of the Nobel Prize money he had won in 1919, with which he tried to set up a business in the porcelain industry. He proved even less successful in commerce than in academia but could not return to his former profession because he had made so many enemies.

After 1933, Hitler put Stark in charge of various scientific bodies, and he conceived grandiose plans for himself as leader of a reorganized German physics, to the dismay of most German physicists. Courageous opposition by Max von Laue (1879–1960)—but not, it has to be said, by Max Planck—saved the Prussian Academy from being taken over by Stark. He seems indeed to have been pathologically aggressive, and he was compelled to retire from his official positions under the Third Reich even before the outbreak of World War II. He lives in memory as a remarkable combination: a Nobel Prize winner in physics with a character that even the Nazis could not stomach.

As long as the Nazi movement remained merely a vocal minority, Lenard and Stark could be dismissed as mavericks who had illegitimately allowed politics to affect their science. But as Hitler rose to power, they became a force to be feared. As we shall now see, even the brilliant Heisenberg, a far greater physicist than either of them, was attacked. However, their odious attempt to intrude racial politics into physics did not survive the pressures of reality, even in Nazi Germany.

Heisenberg's Attitude Toward a German Atomic Bomb

Werner Heisenberg (1901–1976) is famous for his crucial contributions to quantum mechanics during what has been called the "golden age of physics" of the 1920s. He formulated the famous Uncertainty Principle, which states that it is impossible exactly to determine both the position and the momentum of a subatomic particle, from which it follows that predictions can be made only to a certain degree of probability.

As a teenager, Heisenberg studied and argued with the great Niels Bohr and became a favorite pupil of this "pope of physics." He gained a chair in physics at the University of Copenhagen at the age of thirty and a Nobel Prize at thirty-two. In the 1930s, he was at the forefront of research, despite the rejection of the new theories of physics as "Jewish" by the Nazis. When he published a defense of relativity theory in a Nazi journal in 1937, a typically aggressive reply from Stark labeled Heisenberg a "white Jew." Heisenberg appealed to Heinrich Himmler, who was chief of the Gestapo and also a family friend, and after having Heisenberg's loyalties investigated, Himmler protected him against any further such attacks.

In the fateful summer of 1939, when physicists were beginning to realize the possibility of atomic fission, Heisenberg visited the United States and was offered inducements to stay there. But he refused to desert "the nice young physicists" (as he called them) for whom he was responsible in Germany. He was sure that Hitler would lose the coming war, and he wanted to remain in Germany to help preserve what was of value. Once back home, he accepted the directorship of the Kaiser Wilhelm Institute for Physics, then under the control of the Army Weapons Department. This was also the meeting place of the Uranium Society, a group of scientists formed to explore the potential of atomic fission. Heisenberg was thus in a crucial position to influence the course of German wartime research on this topic.

Nuclear fission had been discovered in Germany in 1938 when Otto Hahn (1878–1968), working with Fritz Strassmann, bombarded uranium with slow neutrons. (As early as 1934, Enrico Fermi in Italy and Irene and Frédéric Joliot-Curie in France had produced fission of uranium, but it was not recognized as such.) Austrian physicist Lise Meitner (1878–1968), who worked with Hahn for thirty years until 1938, when she fled to Sweden to escape the Nazis, made the news public. She and her nephew Otto Frisch—who coined the term *fission*—wrote a paper suggesting that the fission of uranium could release large amounts of energy. It was this paper that alerted Bohr (see 9.1) to the danger of a German atomic bomb. Later in the war, the United States became so concerned about the possibility of Germany developing a nuclear weapon that in 1944, it sent a spy to a lecture Heisenberg was giving in Switzerland with instructions to shoot him if he gave any hint that the Nazis were working on such a project! However, despite the fact that the theoretical possibility of an atomic weapon was known to the German physicists, they made no serious effort to construct one.

Why didn't the Germans develop an atomic bomb during World War II? The basic reason seems to be that scientists saw no way, given German resources, to separate the large quantities of the fissionable isotope uranium-235 that Heisenberg thought would be necessary to produce an atomic explosion. So most of their research remained at an academic level and was directed more toward investigating uranium's potential as an energy source than as an explosive. Thus, the Allies' fear of a German atomic bomb turned out to be groundless. Fortunately for the rest of the world, the Nazis had driven out many of Germany's most able scientists when they came to power, and they had so little understanding of science themselves that they did not try to harness it systematically to their war effort.

Heisenberg wrote after the war that because of these factors, he and his colleagues were spared the decision of whether or not to cooperate in developing an atomic bomb for Hitler. He was a German patriot, but he had reservations about the Nazis, especially as it became clear that they were leading Germany to ruinous defeat. It is not clear to what extent Heisenberg realized at the time that an atomic bomb was a practical possibility. Edward Teller later interpreted his not making plutonium to use as fissionable material instead of uranium-235 as a sign of a man working unwillingly. It has been suggested that Heisenberg, while going through the motions of war-related work (and thus saving young physicists from serving in the armed forces), actually tried to divert the minds of the Nazi authorities away from the possibility of an atomic bomb. He himself said after the war:

> Under a dictatorship active resistance can only be practised by those who pretend to collaborate with the regime. Anyone speaking out openly against the system thereby deprives himself of any chance of active resistance. For if he only utters his criticism from time to time in a politically harmless way, his political influence can easily be blocked. . . . If, on the other hand, he really tries to start a political movement, among students for instance, he will naturally finish up a few days later in a concentration camp. Even if he is put to death his martyrdom will in practice never be known, since it will be forbidden to mention his name. . . . I have always . . . been very much ashamed when I think of the people, some of them my friends, who sacrificed their lives on 20 July [in the attempt to assassinate Hitler] and thereby put up a really serious resistance to the regime. But even their example shows that effective resistance can only come from those who pretend to collaborate. (Jungk 1958, 90–91)

Heisenberg here made a utilitarian case for pretence at cooperation with tyranny, while conceding the honorability of those who choose outright opposition. But we have to bear in mind that he must have wanted to put forth the best interpretation of his wartime activities in order to maintain his career in the postwar era.

As the Allied armies invaded Europe after D day, a special scientific unit within the U.S. Army (called the "Alsos") followed the advancing front lines to check possible German progress in atomic fission. They systematically investigated laboratories and rounded up enemy scientists. Heisenberg was one of their last and greatest captures (by then, the fear was that he and his expertise would fall into Soviet hands). He was held with nine other German physicists in a comfortable country house called Farm Hall near Cambridge, England, the catch being that the place was "bugged" so that the Allied intelligence services could listen to the conversations and find out just how much their distinguished prisoners knew. In 1992, transcripts of the conversations recorded in 1945 were finally made public (they had been classified by the British as top secret for almost fifty years). The Germans were not aware that they were being recorded. When the question was raised whether there might be hidden microphones, Heisenberg remarked: "Microphones installed? Oh no, they're not as cute as all that, I don't think they know the real Gestapo methods; they're a bit old-fashioned in that respect" (Bernstein 1992, 49).

Thomas Powers (1993) argued that Heisenberg "killed the project" of building an atomic bomb because he did not want Hitler to have such a destructive weapon. But judging from the recorded conversations, Heisenberg does not seem to have been strongly motivated by any such desire. The central issue is whether Heisenberg's informing the authorities that they would need 100 times as much uranium as we now know is required to build an atomic bomb was due to a scientific mistake on his part or whether he deliberately misinformed the Nazis to prevent them from building an atomic bomb. There is no evidence, apart from his own assertion, that he intentionally misled the German authorities.

When the interned German physicists heard the news of the first U.S. atomic bomb dropped on Japan, they were at first incredulous that the Americans could have overcome the technical obstacles they had themselves thought insuperable, and some of Heisenberg's colleagues dubbed him a "second-rater." He himself was overheard to say, "We wouldn't have had the moral courage to recommend to the government in the spring of 1942 that they should employ 120,000 just for building the thing up" (Bernstein 1992, 51). Once again, this suggests a morally ambiguous state of mind. He apparently felt that his patriotic duty would have been to inform the Nazis of the feasibility of an atomic bomb given sufficient resources, but perhaps he was also somewhat relieved that he had not in fact done so. Ironically, an uncertainty principle seems to apply to Heisenberg's own motives and intentions in the whole affair!

After the war, Heisenberg was allowed to return to his native land, and he took his place as the leading physicist of democratic West Germany, becoming director of the Physics Institute of the newly renamed Max Planck Gesellschaft from 1946 to 1958. Like many other physicists, he explored philosophical issues in later life, and his Gifford lectures given at St. Andrews University in Scotland in 1955 and 1956 were published under the title *Physics and Philosophy: The Revolution in Modern Science.* He had regained academic respectability.

8.2 Scientists and Communism in the USSR

Before we look at some particular episodes, let us summarize briefly the story of science in the USSR after the October Revolution of 1917. Scientists had been among the well-off elite under the Czars, and most of them, though supporting democratic reform, had opposed the more extreme Bolshevik faction. They therefore became objects of suspicion after the revolution when the Bolsheviks took power, and during the civil war of 1918 to 1921, many of them fled abroad. But once the new regime was firmly established, Vladimir Lenin realized the need for technical expertise in developing the country, and over the next decade many talented scientists were installed in newly formed institutes housed in former aristocratic palaces, given funding to pursue their work, and even allotted special rations in times of shortage. (Lenin was a strong believer in technology—during the 1920s, he used the motto "Socialism and electrification!") There followed a dramatic acceleration in the pace of scientific education and research. Practical benefits for the new socialist society were expected, of course, but they were not demanded in the short term, and at first the government exerted little ideological pressure on scientists.

During this period, a Physical-Technical Institute was established in Leningrad; it was the foundation of the remarkable Soviet achievements in nuclear physics in later years. Other institutes were set up for the study of optics and radium. Botanist N. I. Vavilov (1887–1943) was invited to set up a Laboratory of Applied Botany in a palace in the center of Leningrad (which is now St. Petersburg once again), where he began assembling what was to become the world's greatest collection of plant species (as a genetic pool for selection and hybridization). Some Soviet scientists became world leaders in their fields: geneticist S. S. Chetverikov (1880–1959) laid the foundations of population genetics; physiological psychologist I. P. Pavlov (1849–1936) did his famous work on conditioned reflexes; and biochemist A. I. Oparin (1894–1980) put forward the first scientific theory of the origin of life on earth.

But from about 1929, as Stalin tightened his dictatorial grip on power and a younger generation of Soviet-trained scientists sought advancement, the early golden years of Soviet science came to an end. Rapid industrialization and forced collectivization of agriculture were the order of the day. Short-term practical results were now expected of loyal Communist scientists. Accusations of sabotage and disloyalty were common—perhaps to distract attention from failures in Stalin's ambitious programs. Anyone who could be labeled a "bourgeois intellectual" was especially under suspicion, and people began to "disappear." In 1929, the USSR Academy of Sciences was declared to be a center for counterrevolutionary activity, and three of its members were arrested. In this atmosphere of terror, scientific arguments easily became politicized, and it was possible for incompetent would-be scientists to invoke political power against their colleagues. For example, when disputes arose in agriculture and biology, Lysenko was able to rise to scientifically undeserved prominence (as we are about to see). A number of distinguished scientists ended up on the wrong side politically; Chetverikov, for example, was exiled to the Urals and was not released until he was a very old man in 1955. Vavilov was arrested in 1940 and died in a labor camp.

The Stalinist dogma of the time was that "bourgeois science" (by which was meant the science developed in the capitalist West), though it had its successes, was limited in scope, supposedly because capitalism restricted science and technology to a certain level of development. The hope was that Soviet science would first catch up with this level and then transcend it by developing that further potential as only communism could. But even after World War II, many sciences (with one exception being physics) made little progress in the USSR, for reasons that we shall now illustrate.

How Lysenko Got Stalin's Support for Pseudobiology

Russian biologist (or as many would rather say, pseudobiologist) Trofim Denisovich Lysenko (1898–1976) attained the very peak of scientific administrative power under Stalin and was able to suppress much genuine biological research, especially as related to genetics, for a whole generation in the Soviet Union. Lysenko was only marginally a scientist. Educated in applied agronomy, he pursued a career in scientific administration without making any significant contributions to scientific research. But his story is instructive in showing how wrong things can go in science under political conditions less tolerant than those we normally take for granted.

Lysenko appears from the start to have had a greater talent for self-publicity and political infighting than for science itself. He got a heroic-sounding feature story about himself published in *Pravda* in 1927, based on experiments on winter planting of peas to precede the cotton crop in Azerbaijan. The journalist wrote that Lysenko had "solved the problems of fertilizing the fields without fertilizers and minerals" but went on to describe him in less complimentary terms as a man with "a dejected mien" who was "stingy of words and insignificant of face." "Only once," it was noted, "did this barefoot scientist let a smile pass, and that was at mention of Poltava cherry dumplings with sugar and sour cream."

The "barefoot" professor gained followers, pupils, and a field in which to conduct experiments. He was photographed being visited in the winter by agronomic luminaries who stood before the green fields of the experimental station, gratefully shaking his hand. The winter crop of peas apparently did not prove a success in subsequent years, but it helped Lysenko get a strong foothold on the ladder of power. In his first major scientific paper, he tried to correlate the time and heat that a plant needs to grow through all its phases, from sprouting to production of new seed. But he took almost no account of previous investigations of the topic and also committed an elementary error in statistical reasoning. Biologist N. A. Maksimov politely put him right, but Lysenko took no proper notice of the criticism, merely dropping any attempt at statistical argument and dogmatically asserting that the germinating seed's being chilled for a certain period of time is the only determining factor in the maturing of winter grain.

In 1929, Lysenko acquired a wider reputation in the USSR for his alleged discovery of "vernalization," which was in fact a traditional practice of hastening the germination of seeds by immersing them in water and then freezing them. He put a new spin on the process by claiming that it would also cause future generations of seeds to germinate earlier. He persuaded his Ukrainian peasant father to soak his winter wheat seeds and bury them in a snowdrift before sowing them in the spring. This particular crop, according to the Ukrainian Commissariat for Agriculture, did better than the spring wheat sown in neighboring fields. Yet even before this test, stories of sensational progress in farming had been handed out to the press. Lysenko proposed to grow tomatoes in the tundras and sugar beets in the deserts of Central Asia. In October 1929, Lysenko was promoted to the All-Union Institute of Plant Breeding in Odessa. He was on his way up, and scientists like Maksimov now had to express respect for his claimed practical achievements, however skeptical they may really have been.

In the 1930s, Lysenko became director of the Odessa Institute and was able to found his own scientific journal. He extended the idea of vernalization into a theory of "phasic development" of plants and put forward a more general theory of heredity that focused on the whole organism rather than on the genes in germ cells, as in neo-Mendelian genetics. Since such speculations were formulated in a way too vague to be tested, they hardly count as genuine scientific hypotheses. But Lysenko was now able to use his political influence to frighten off his critics. At the time of the collectivization of agriculture, when the *kulaks* (comparatively wealthy peasants) were being forced off their land, he included the following denunciation in a speech to the Second All-Union Congress of Shock Collective Farmers in 1935: "Comrades, kulak-wreckers occur not only in your collective farm life. You know them very well. But they are no less dangerous, no less sworn enemies also in science. No little blood was spilled in the defense of vernalization in the various debates with some so-called scientists. . . . A class enemy is always a class enemy whether he is a scientist or not." After these sinister words, Stalin is reported to have said, "Bravo, comrade Lysenko, bravo!"

Once Lysenko gained the backing of Stalin, anyone advocating views in biology different from his own became politically suspect. During the great purges of the late 1930s, Soviet scientists who were suspected of disloyalty could find themselves suddenly arrested, interrogated, sent to labor camps, or even executed. Maksimov, Lysenko's early critic, was in trouble; even the internationally recognized Vavilov, director of a prominent botanical institute since the early years of the USSR, was arrested in 1940 and died of malnutrition in a Siberian labor camp in 1943.

Lysenko was appointed director of the Institute of Genetics of the USSR Academy of Sciences in 1940. However, it was not until 1948, well after the Stalinist reign of terror and the "Great Patriotic War," that he attained complete control over Soviet biology. Just prior to that time, it had looked as if the practical failures of his much-advertised agronomic measures would undermine his position, but again he managed to get support where it mattered. At a meeting in 1948, Lysenko delivered a speech edited by Stalin himself in which he was able to silence all opposition by announcing that his own "report" had the support of the Central Committee of the Communist Party. The gist of this notorious report was that there were two opposing schools of biological thought in the USSR: one that was materialistic and Soviet and another that was idealistic, reactionary, and committed to a chromosome-based theory of heredity. Lysenko associated himself with the "ideologically correct" views of I. V. Michurin (1855–1935), a breeder of fruit trees who believed in the inheritance of acquired characteristics. The Mendelian view became officially anathema, to be repressed with all the force of the Soviet state.

Swiftly thereafter, a campaign ensued to dismiss all scientists associated with the condemned genetic theories (which were already a standard part of biology in the rest of the world). Journals and textbooks were purged of genetics, and an order was even issued to destroy stocks of *Drosophila*, the fruit fly so often used in genetic experiments. Meanwhile, Lysenko's supporters were appointed to key positions, so any scientists who believed in Mendelian genetics had to pay lip service to Lysenkoism to retain their posts. He was at the peak of his power, but his position began to weaken even before Stalin's death in 1953. In 1950, Stalin, in striking contradiction to his previous policy, announced that "no science can develop and prosper without the clash of opinions, without freedom of criticism"; in 1952, criticism of Lysenko's theories duly began to appear in print again. Since he had been in charge of all Soviet agronomy for a number of years, the practical failures of his policies could not forever be concealed. In the "Great Stalin Plan for the Transformation of Nature," for example, millions of trees had been planted, according to Lysenko's direction, in clusters rather than singly. The theory was that they would effectively thin themselves out. Careful assessment of the actual results (something new in Soviet agronomy) showed that more than half the trees had died.

Mendelian genetic theory began slowly and cautiously to revive alongside Lysenkoism. Yet it was not until after Nikita Khrushchev lost power in 1964 that Lysenko was dismissed from his directorship of the Institute of Genetics and his journal *Agrobiology* was discontinued. He was allowed to keep the substantial experimental farm at Lenin Hills until his death. Since then, Lysenkoism has disappeared almost without a trace from science in the Soviet Union. It had some influence for a time in Eastern Europe and in Maoist China, but it never established a permanent hold anywhere.

Lysenko's desire for fame and power seems to have overwhelmed any scientific curiosity to seek the truth. In striking contrast to Mendel, who sought discussion of his work but was starved of such stimulus, Lysenko failed to respond to expert criticism. We may choose to call individuals like Lysenko, who have only a smattering of scientific knowledge and who merely go through the motions of experimental testing, bad scientists—or we may decide that they are not really scientists at all. His achievement was to retard Soviet genetics for a generation and handicap the development of more efficient agricultural production.

How, then, did it happen? An adequate account must go beyond the personal qualities of Lysenko or even of Stalin. It may be tempting to try to explain Lysenko's attaining power over the scientific community by appeal to the Marxist ideology that officially dominated all Soviet thought and practice. Many writers have noted that some of his speculations echoed the pre-Darwinian theory of Lamarck (1744–1829), who proposed that characteristics acquired during the lifetime of an individual organism could be inherited. And it is plausible to see Lamarckism as congenial to Marxist ideology, since it suggests that human nature can be altered by socioeconomic conditions, unlike Darwinian theory, which holds that environmental influences on an organism do not affect the genes passed on to its offspring. But considerations at this level of theoretical biology or political philosophy were not what motivated Lysenko himself. He clearly had little understanding of biological theory and even admitted to Stalin that he had never studied Darwin's work. He was primarily a practical agriculturalist, whose rise to great power over the scientific community was made possible by the nature of the Stalinist regime.

The Soviet Union had encouraged an elite of theoretical scientists, some of whom were world renowned, like Vavilov, Chetverikov, Pavlov, and Oparin. But the political leadership was increasingly impatient with the lack of immediate practical use for scientific investment, especially in the vital area of agriculture. This situation provided an opportunity for a new generation of ambitious young people to win promotion by claiming to have ways of increasing agricultural production quickly. It seems to have been this political and economic background, more than any direct influence from Marxist ideology, that made the whole Lysenko affair possible.

In Lysenko we are presented with an ill-educated but ambitious and somewhat unscrupulous would-be scientist who, given the peculiar political circumstances of his time, was able for a while to exercise dictatorial power over much of the scientific community of a huge state. As long as he retained his political position, he was able to suppress teaching and research in the whole field of genetics. Such ideological interference in the very content of scientific thought is an aberration from what we think of as normal practice. It has been assumed to be fairly rare in the history of science: the cases of Galileo and the Catholic Church (see 7.1) and the Nazi rejection of Einstein's physics (see 8.1) are usually cited as exceptional. But we had better be prepared to acknowledge that social and political factors can also affect scientific belief in more democratic contexts, especially where the subject matter involves deeply ingrained convictions (see the discussion of Burt on intelligence in 5.2). Even if unusual, Lysenko's case is instructive; in science, as elsewhere, we learn a great deal about how something works by examining how it can go wrong.

Kurchatow's Loyal Service to the Soviet State

Throughout the Stalinist years of terror and dogma, physics in the USSR remained relatively unscathed. Realizing the vital military importance of physicists, Stalin excused them from the "political education" that was compulsory for all other scientists. Igor Kurchatow (1903–1960) was a member of the first generation of Soviet-trained scientists, having graduated from the University of the Crimea in 1923. Four years later he moved to the Leningrad Physical-Technical Institute and went on to become head of its nuclear physics department. He and his team built the first cyclotron (particle accelerator) in the world, and like physicists in other countries, they realized the possibility of chain reaction and nuclear fission in uranium. When Hitler attacked the Soviet Union in 1941, all available scientific resources were thrown into war production. Often this meant using the expertise of people who had been interned in labor camps, and a system of "prison laboratories" and aircraft factories grew up that was to outlast the war. A large laboratory was established under Kurchatow's supervision in Moscow, and it was later named the I. V. Kurchatow Institute of Atomic Energy in his honor.

After the United States used atomic bombs at Hiroshima and Nagasaki in 1945, Stalin had to acknowledge the success of "capitalist" science. He ordered Kurchatow and his team to work flat out on a Soviet bomb. By the end of 1946, they had constructed the first atomic reactor in Europe with the aid of prison labor, some captured German scientists, and important espionage data from the West (notably via Klaus Fuchs, who had acted as a spy inside the U.S. atomic bomb project). To the consternation of the West, the first Soviet bomb was exploded as early as 1949; with the crucial contribution of Andrei Sakharov (discussed in the following section), they tested their first hydrogen bomb in 1953.

Kurchatow also did much for the civil use of atomic energy and was instrumental in the construction of the USSR's first nuclear power station in 1954. The very rapid Soviet development of this energy source, whose human and ecological dangers have been all too clearly demonstrated by the explosion at Chernobyl, owes much to Kurchatow. Obviously, his scientific expertise and organizational ability was equaled by his loyalty to the state; he was three times honored as a Hero of Socialist Labor, and his ashes were entombed in the Kremlin wall. But he remains a grey figure of whom almost nothing is known beyond his scientific distinction and his political subservience. There are many more of his type in every country.

Sakharov's Courageous Dissidence

The life of Andrei Sakharov (1921–1989) shows how in a totalitarian society—as the Soviet Union was until the last year or two of his life—it can sometimes be possible for scientists to use the special position their expertise gives them to exert an unusual degree of moral and political influence.

As we have seen, physics remained relatively untouched by the excesses of Stalinism. By 1949 Soviet scientists were able to produce an atomic bomb. Sakharov completed his doctorate with extraordinary speed (by Russian standards) and was set to work in 1948 to develop a thermonuclear (hydrogen) bomb. His technical success in this project was such that for a short time the Soviets had an advantage over the United States, because Sakharov had seen how to make an H-bomb of manageable size, using a fission-fusion-fission technique, while all the Americans had was a device too huge to get on an airplane. Thus, the USSR rapidly caught up with the United States in the nuclear arms race. In reward, Sakharov was elected a full member of the Academy of Sciences in 1953, at the unprecedented age of thirty-two.

Reaching that position just as Stalin died gave him a unique opportunity. He was privileged materially, of course, but also politically in that the vital national importance of his expertise in nuclear physics gave him a certain measure of independence, a freedom to speak out that hardly anyone else in the USSR enjoyed. He soon realized how poisonous was the radioactive contamination produced by nuclear bomb tests in the atmosphere, and he pleaded with the Soviet leaders to desist from them. When a three-year moratorium on such tests expired in 1961, he personally lobbied Khrushchev not to resume them, only to be told in no uncertain terms that he should stick to science and let Khrushchev handle the politics. He also took part in the scientists' campaign to free biology from Lysenko's still-persisting dictatorship.

Sakharov's concerns soon widened from scientific to social and political matters. He signed a collective letter warning against any attempt to rehabilitate Stalin. He became a public defender of free thought and human rights and developed increasingly close contacts with other dissidents. In 1968, he circulated within the USSR—and had published in the West—a remarkable manifesto entitled "Progress, Coexistence and Intellectual Freedom," which suggested how the two superpowers could cooperate to solve the world's problems. After this, he was denied security clearance and was dismissed from his research position. Thus, by force of official policy, he ceased to be a working scientist.

All throughout the 1970s he turned to the West for publicity about many individual injustices in the USSR. In 1975, he was awarded the Nobel Peace Prize, but he was not permitted to travel to receive it. He was warned to stop meeting with anti-Soviet forces, and a period of official persecution began. He and his family were subjected to all manner of harassment by the KGB, and in 1980 he was arrested and sent into internal exile in Gorky. There he undertook hunger strikes several times in an attempt to induce the authorities to allow his wife to go to the West for urgent medical treatment. For this act of defiance, he was subjected to the torture of force-feeding (the authorities did not want to have his death on their hands). Support from the West, especially from some American physicists, was important in discouraging the worst excesses of the KGB and encouraging Sakharov's eventual release. When Mikhail Gorbachev came to power in 1985, the system at last began to relent, and Sakharov's wife Elena Bonner was allowed to go abroad, although he himself had to stay because he had had access (some years before) to military secrets.

Under Gorbachev's new policy of *glasnost*, Sakharov was released from internal exile, and he immediately took a leading place amongst those campaigning for further reform. As someone who had for so long made eloquent statements for freedom of thought, human rights, peaceful coexistence, and care for the environment and who had courageously endured the worst that the system could do to him, he had tremendous moral authority. He soon found a place in the new, open, and genuinely lively parliament and immediately demanded even more rapid change toward democracy. His national stature had become such that Gorbachev himself had to pay his respects at Sakharov's funeral in late 1989. No doubt he could have made more of a contribution to pure physics had he not chosen to involve himself so deeply in social matters or had he been born into a different political system in which such commitment was not so urgently needed. But perhaps his moral and political achievement in the USSR will be judged as being at least as important as the scientific work he might otherwise have done.

8.3 Scientists in Contemporary Totalitarian Regimes

The choices for scientists under present-day dictatorships remain stark. Consider the case of Hussain Shahristani, an Iraqi nuclear physicist, educated in Britain and Canada, who refused to cooperate in building an atomic bomb for Saddam Hussein. His expertise would have been crucial to such a project, and if he had participated, Iraq might well have had a nuclear weapon to use in the brief Gulf War of 1991. Shahristani was called in by one of Saddam's lieutenants and told, menacingly, "Anybody who does not serve his country does not deserve to be alive, so I hope you understand the message." He bravely replied, "Yes I know what you mean, perhaps I even agree with you that we are all obliged to serve our country, but I may have a different view of what constitutes service" (Bhatia 1992). In fact, Shahristani's brains were of too much potential value to the regime for him to be killed. He was hideously and repeatedly tortured, and threats were made against his wife and three small children. When he heroically maintained his refusal to serve in the Iraqi nuclear program, he was sentenced to twenty years' imprisonment. He served ten years—much of it in solitary confinement—but was able to escape across the border to Iran in the confusion caused by the bombing of Baghdad during the Gulf War. Other Iraqi scientists, to Shahristani's disgust, cooperated with Saddam's regime quite willingly or gave in under pressure. But one would have to apply a very high moral standard indeed to criticize anyone for capitulating to such barbaric and ruthless methods.

In other countries, scientists who publicly dissent from the ruling regime also get into major trouble. In China, astrophysicist Fang Lizhi, together with his wife Li Shuxian (also a physicist), spoke out against the ruling Communist party at various times since the 1950s. At the time of the incipient democracy movement of the 1980s, Fang Lizhi was one of its intellectual leaders (some Chinese called him "our Sakharov"). When the movement was brutally crushed by tanks in Tiananmen Square, he and his wife took refuge in the U.S. embassy in Beijing and were later able to emigrate to the West. In this case, it was not so much the content of his area of expertise as it was his being a distinguished scientist with Western connections that gave Lizhi the opportunity to be a leader of political opposition. The independent-minded, critical spirit required for the proper practice of science (see 1.1 and 3.1) does not easily fit into totalitarian societies. But it is quite common for scientists (like many other specialists) to keep their work in one mental compartment, separate from the other aspects of their lives, whether personal, political, or religious. Under harsh and brutal regimes, one can hardly blame scientists for getting on with their work, prudently keeping their heads down while hoping for better times. It is only the dramatic and heroic cases of conflict and opposition that we get to hear about.

Suggestions for Further Reading

Bernstein, J. 1992. "The Farm Hall Transcripts: The German Scientists and the Bomb." *The New York Review.*

Cassidy, D. C. 1992. *Uncertainty: The Life and Science of Werner Heisenberg.*

Heilbron, J. 1986. *The Dilemmas of an Upright Man: Max Planck as a Spokesman for German Science.*

Joravsky, D. 1970. *The Lysenko Affair.*

Jungk, R. 1958. *Brighter Than a Thousand Suns: A Personal History of the Atomic Scientists.*

Medvedev, Z. 1969. *The Rise and Fall of T. D. Lysenko.*

Powers, T. 1993. *Heisenberg's Wars: The Secret History of the German Bomb.*

Sakharov, A. 1990. *Memoirs.*

9 Scientists and Public Policy

Not many scientists attain political power, and in the rare cases in which one does, he or she inevitably ceases to be a working scientist. No one has enough time and energy to make a success of both occupations, each being so demanding in its own way. One example of this trade-off is Chaim Weizmann (1874–1952), the chemist who aided the British war effort as director of the Admiralty Labs from 1914 to 1918 and extracted in return a promise from the British government to support the establishment of a homeland for the Jews. He then gave up science and devoted himself very single-mindedly and successfully to the Zionist campaign to found the state of Israel, of which he became the first president. Very few scientists, however, exercise any significant political influence, and scientists will be found to take as great a variety of attitudes toward the government and the political system of their respective countries as other citizens. We will be concerned here only with cases in which there is some relationship between a scientist's expertise and political involvements.

In the previous chapter, we illustrated the options open to scientists under totalitarian regimes. We will now look at some instructive case studies of the involvement of scientists in democratic politics. A few have exercised considerable behind-the-scenes influence on government policy. It is, of course, their knowledge of some relevant field—rockets, bombs, nuclear energy, agriculture, or medicine—that qualifies them as expert advisers. But technical expertise is not enough; to be influential, a scientist needs to have the ear of her political masters, and she is unlikely to get it unless she is in basic sympathy with their aims (or at least presents herself as such).

Most scientists are primarily concerned with their own research and will be ready to do anything (anything legal, anyway!) to get the means and the freedom to proceed with it. Scientists may campaign vigorously for better public funding, they may argue behind the scenes for it, and they may lobby against governmental restrictions on their projects. As well, scientists may have strong views about how their specialized knowledge should be used and may lobby to turn those views into policy. Biologists and ecologists tend to have policy concerns connected with their own topics of study, for example, halting pollution of the environment and the destruction of species and ecosystems, and many scientists have publicly committed themselves to such causes.

9.1 Scientists and Nuclear Weapons

We shall now examine the role of certain physicists in what is perhaps the most famous (or notorious) of all applications of science—the construction of the first atomic bomb in the United States in the early 1940s and the postwar development of increasingly sophisticated nuclear weapons.

First, a brief sketch of the scientific background is needed. From the beginning of the twentieth century, the work of Joseph Thomson, Ernest Rutherford, and Niels Bohr started to unravel the internal structure of the atom. In the 1920s, classical physics was rapidly restructured by the development of quantum mechanics. An amazingly gifted generation of theoretical physicists contributed to this intellectual revolution, whose consequences still reverberate around the world. There were three main centers of fundamental physics research at the time, all of them in Europe—Rutherford's Cavendish Laboratory in Cambridge, Bohr's Institute of Theoretical Physics in Copenhagen, and a many-talented group in Goettingen in Germany, which included the young Heisenberg (see 8.1). Applications for their work were not foreseen at this stage—it was pure science, done for its intellectual interest alone. Lord Rutherford maintained until his death in 1937 that talk of releasing atomic energy was "moonshine." American physicist A. O. C. Nier, who first developed the technique used to isolate uranium-235, the isotope that fueled the first atomic bomb, recalled in an interview that no one in government took basic atomic research very seriously in the early 1930s. The physicists were "interested in finding out what made things tick inside the nuclei of atoms" (Nier 1978, 3). Nier noted that in order to get financing to build accelerators in those days, they had to appeal to possible medical applications.

Szilard: Worries About Nuclear Warfare

Leo Szilard (1898–1964) was a Hungarian-born physicist who had to change countries three times, first to Germany to escape the upheavals in his native land and get a scientific education in Berlin, then to England to escape the rising tide of Nazism, and finally to the United States to find employment. In 1932, Sir James Chadwick (1891–1974) discovered the neutron, and it was suggested that such a particle without electric charge might be able to penetrate the atomic nucleus. In 1933, just after his flight to England, Szilard seems to have been the first scientist to conceive of the possibility of a nuclear chain reaction if an element whose nucleus would emit two neutrons after swallowing one could be found. Realizing the enormous energy that might thus be released and being all too vividly aware of the darkening political situation in Europe (he had formed strong antimilitarist views after his early experiences in Hungary during World War I), Szilard suggested to physicists that they should stop publishing their results to keep them secret from potential enemies. But at this stage, before any real prospect of atomic energy was apparent and before the Nazis became a threat to their neighbors, nobody took seriously this unusual proposal for restricting the free flow of scientific ideas.

In fact, atoms were being split in various labs without anyone realizing it. The first chain reaction was probably brought about in Rome by Enrico Fermi (1901–1954). In 1935, Frédéric and Irene Joliot-Curie received the Nobel Prize for their discovery of artificial radioactivity. In their speech at Stockholm, they suggested that scientists might now be able to cause nuclear transformations of an enormously explosive character. Yet even these dramatic words aroused only passing interest. In 1938, Hahn and Strassmann, working in Nazi Germany, had demonstrated that what had been happening in Fermi's experiments was the "bursting," or fission, of the nucleus of uranium. Early in the following year, Otto Frisch and Lise Meitner (a rare example of a woman who gained recognition on the frontier of physics) showed that large amounts of energy were liberated by such fission. (See the discussion of Heisenberg in 8.1.) These papers were quickly published, in accordance with standard scientific practice. Szilard (by then in the United States) confirmed to his own satisfaction that further neutrons were given off in the fission of uranium, so the chain reaction he had envisioned really could happen. Again he suggested a voluntary moratorium on publication, and this time he was listened to, at least in the United States; physics had progressed, and politics had degenerated, since his previous proposal. But as is typical at the moving frontier of research, other scientists, in this case Frédéric Joliot-Curie of France (son-in-law of Marie Curie), were on the point of making the same discoveries themselves. Displaying normal scientific keenness to take the credit (as well as French unwillingness to take advice from abroad), Joliot published the result in *Nature* in April 1939.

It seemed as if physics was presenting the world with a huge, new, and explosive force just in time for the war that now appeared inevitable. German physicists clearly knew about the theoretical possibility of nuclear fission, and when Germany forbade the export of all uranium ore, it looked as if the Nazi government knew, too. If Germany were to acquire sole use of atomic bombs, it might rule the world. Szilard and his colleagues (including Fermi and the youthful Edward Teller) were now urgently concerned with persuading the U.S. government to take the threat seriously. But refugee scientists who were not even U.S. citizens had no political clout in Washington and no means of access to the highest authorities. The difficulty was eventually overcome by an introduction to President Roosevelt's friend and adviser Alexander Sachs, who encouraged Szilard to draft a letter to the president. As a sort of "halo" on this fateful missive, Szilard and Teller persuaded Einstein (who enjoyed legendary scientific fame) to add his signature. Sachs finally succeeded in bringing it to Roosevelt's attention in October 1939, a month after the war began in Europe. Two years of experimentation and bureaucratic discussion followed before the U.S. government eventually decided to try to develop an atomic bomb. Szilard's energetic persistence in bringing the new atomic potential to the notice of the strongest power in the world had at last borne fruit in a major governmental decision.

Szilard then contributed to the success of the "Manhattan Project" (the code name for the U.S. effort to develop an atomic bomb). He and Fermi brought about the world's first controlled nuclear chain reaction in former squash courts at the University of Chicago in December 1942. It was not long before Szilard became concerned about a new political problem, namely, the use that the United States might make of the weapon. As the production of the first atomic bomb neared completion, Germany was already defeated by conventional means, and U.S. authorities began to consider using the new bomb to finish the war against Japan. Szilard said afterward that "in 1945, when we ceased worrying about what the Germans might do to us, we began to worry about what the government of the United States might do to other countries" (Rouzé 1963, 70). Once again he got Einstein's signature on a letter to Roosevelt, this time warning against an atomic arms race, but the ailing president did not get to read the letter before he died. Szilard then tried to reach the new president, Harry Truman, but was only given an interview with James F. Byrnes (who later became Truman's secretary of state). Ever the far-sighted political thinker, Szilard laid before Byrnes his ideas for international control of atomic energy, involving partial surrender of national sovereignty to the extent of allowing Soviet inspection of U.S. installations and vice versa. Only very recently have any such arrangements been seriously considered, but in an America still in the closing stages of a costly war and before the first explosion of an atomic bomb, they must have seemed utterly fantastic to an average politician like Byrnes.

Szilard discussed the possible uses of the bomb with his fellow workers on the Manhattan Project, and their worries reached the University of Chicago, which appointed a committee of seven scientists to discuss the "social and political consequences of atomic energy." Its chairman was physicist James Franck, but Szilard was also a member, and it seems that his ideas strongly influenced the eloquently argued document that came to be known as the "Franck Report." With great prescience, they wrote, "If no efficient international agreement is achieved, the race for nuclear armaments will be on in earnest not later than the morning after our first demonstration of the existence of nuclear weapons." The committee advised against their immediate military use against Japan and recommended instead that the first explosion be on a desert or a barren island, with U.N. observers, so that the United States could say to the world: "You see what sort of a weapon we had but did not use. We are ready to renounce its use in the future if other nations join us in this renunciation and agree to the establishment of an efficient control." This report was submitted to the U.S. government, which referred it in June 1945 to a panel of four atomic experts. The panel did not support the idea of a demonstration explosion, and in August the first two bombs were dropped on Hiroshima and Nagasaki.

After the war, Szilard continued to try to influence political policy and was a leader of the initially successful lobby for civilian control over atomic energy. But in most matters he had little influence. It was unlikely that the U.S. authorities were going to take advice from scientists on matters of political and military policy. The genie of atomic power was truly out of the bottle.

Bohr: Failure to Influence Churchill or Roosevelt

Niels Bohr (1885–1962) was one of the greatest theoretical physicists of this century. Perhaps more than anyone else, he can be credited with unveiling the mysterious inner secrets of atomic structure. Toward the end of World War II, as the Anglo-American atomic bomb project neared its fruition, Bohr's scientific preeminence briefly won him the ear of the leaders of both countries. But as we shall see, he was no more able than Szilard to influence them with his fine ideas about postwar international cooperation.

Bohr combined love of his own small country, Denmark, with receptiveness to other cultures and currents of thought. After earning his doctorate at Copenhagen, he went to England to study with J. J. Thomson at Cambridge and with Ernest Rutherford, who was then at Manchester.

Back in Copenhagen as an assistant professor before World War I, the brilliant young Dane was already beginning to revolutionize his subject. He took up Planck's idea that energy comes in discrete indivisible packets and applied it to atomic structure, suggesting that an atom can pass abruptly between a number of possible states, emitting or absorbing a quantum of energy in each transition. After the war, Bohr was given a full professorship, and a new Institute of Theoretical Physics was built for him, which was to be his intellectual home for the rest of his life. He was awarded the Nobel Prize in 1922. Through the following decade— those golden years of physics when knowledge of the field moved so far and so fast, with a galaxy of talents making variously brilliant contributions—Bohr and his institute were pivotal to this quickly developing subject. With his impish humor, his apparent inability to state anything clearly, and his unique intuition about tiny but powerful atomic realities, he teased out the meaning and validity of the paradoxical new theories.

By 1939, it was clear that release of atomic energy was possible in a chain reaction involving fission of uranium atoms. Bohr realized that the only problems remaining were technical. Despite tempting offers from the United States, he felt he had to stay in his native land and do what he could to keep his institute going. So he did not leave when the Nazis invaded in 1940. But as the tide of war began to turn against Germany, Danish resistance increased, and the Nazis started arresting leading citizens. Bohr escaped in a risky nocturnal passage to Sweden in a small boat. In Stockholm, a telegram soon came to him from Lord Cherwell (Churchill's advisor on science) with an invitation to England, the obvious but unstated implication being that he was needed to help with atomic work. A British plane flew to neutral Sweden to pick up this precious cargo of nuclear expertise. Bohr had to sit in the bomb bay, equipped with a parachute and flares in case of emergency ejection. The plane flew high, and Bohr was told to turn on his oxygen supply, but the helmet did not fit his huge head; he failed to hear the instruction and lapsed into unconsciousness.

Mercifully, that unique scientific brain was unimpaired and soon at work in London, advising on "Tube Alloys," the code name for the British atomic bomb project. Even at this early stage, Bohr was raising worries he had about a postwar atomic arms race. Rivalry already existed between the British and American teams, and they competed to recruit Bohr. But after agreement on full collaboration, the British consented to send all their top physicists to the United States, and thus it was that Bohr crossed the Atlantic. He was met by the U.S. Secret Service and instructed that he must call himself "Nicholas Baker" in the future, something he could not always remember to do. Bohr spent the summer of 1944 in Los Alamos, New Mexico, where a military laboratory had been specially constructed for the nuclear project because it was best for both scientific and security reasons to have all the research conducted at one location. There he found many old scientific friends hard at work, and it was decided that the best way to use Bohr's unique expertise would be to have him make a general review of all phases of the work to ensure nothing had been overlooked.

Back in Washington, Bohr found a chance to air his concerns about the postwar situation as an advisor to President Roosevelt. The president expressed interest, being aware that nuclear proliferation was a problem that he and British Prime Minister Winston Churchill would have to face sooner or later. Bohr got the impression that he was authorized to tell Churchill that the president would welcome suggestions on the matter. Thus, he was sent back to London believing that he had a chance to persuade the Allied leaders to internationalize all use of atomic energy after the war and avoid the arms race between East and West that he feared.

Unfortunately, the interview with Churchill was a disaster. Bohr was not noted for his clarity even to physicists working in his own field (some have reported that it was often difficult to decide what language he was trying to speak!). Upon being ushered in to meet the bulldoggish prime minister, Bohr began to speak in his low, whispering voice, building up his argument slowly as in a seminar, but Churchill had no patience for any such academic performance. When Lord Cherwell tried to put in a helpful word, he and Churchill got into an argument of their own, which took up much of the half hour allotted for the interview. Bohr did not even get as far as his main point, that the new atomic knowledge forced the choice between a dangerous arms race or a new international order. Despairingly, he asked on his way out if he could send Churchill a memorandum of what he had wanted to say, but the latter's reply was, "It will be an honour for me to receive a letter from you—but not about politics." It is probable that nothing even the most eloquent lobbyist could have said would have made much difference. Churchill was resolutely opposed to any breach of the secrecy surrounding the atomic bomb, especially as regarded the Soviets.

Back in Washington and crestfallen, Bohr had a second chance when President Roosevelt expressed an interest in hearing his ideas. Accordingly, he entered the Oval Office, armed with a carefully drafted memorandum. This time he met with what seemed a much better reception; the relaxed, genial Roosevelt joked about Churchill's belligerent moods and made some show of support for international control after the war. (The president was known for his charm and his tendency to let people think he agreed with them.) Yet when Roosevelt and Churchill met again in 1944, any sympathy the former might have had with Bohr's proposals disappeared like smoke in the wind. The two leaders agreed to maintain the utmost secrecy about the incipient atomic bomb, discussed its use against Japan, and proposed that Anglo-American work on it would continue after the end of the war. Churchill even cast doubt on Bohr's reliability and loyalty and accused him of breaking secrecy about the bomb. His advisors had to argue very strenuously that any such suspicions were completely unjustified.

Why did Roosevelt not follow up on his apparent sympathy with Bohr's ideas? There are several plausible reasons: He was getting physically weaker (he was to die in less than a year), perhaps he did not like to disagree with Churchill, and perhaps his distrust of Stalin was increasing. But very strong political factors were also at work, which persist to the present day. No nation likes to abandon an apparent advantage over rivals, as the proposal for sharing atomic knowledge would have implied. Nor does any state willingly give up any aspect of its sovereignty, as the enforcement of international control would have required. No leader seems able to resist for long the feeling that his country must be as strong as possible.

Oppenheimer: Rise and Fall from Political Grace

J. Robert Oppenheimer (1903–1967) came to center stage with the construction of the first atomic bomb. He was the scientific director of the Manhattan Project, which, by means of an unprecedented concentration of scientific and industrial resources, succeeded in taking a theoretical possibility and turning it into a usable weapon. But despite the public adulation that this achievement brought him and the influence on U.S. atomic policy that he exercised for a time after the war, he was to fall from power in highly controversial circumstances.

The precocious young Robert Oppenheimer majored in chemistry at Harvard while developing a deeper interest in physics. He went to England and then to Goettingen in Germany in the 1920s, where he found the cosmopolitan atmosphere and the ferment of ideas more congenial. He completed a Ph.D. and began to feel that he could make a real contribution to physics. On his return to the United States, he took up a professorship at the University of California at Berkeley, where he built up an important school of theoretical physics from almost nothing during the 1930s. He was a hero to his students, at least to those who became members of his clique. Through friends like Jean Tatlock, he became involved in left-wing activities. At this time of sympathy with the Republican side in the Spanish civil war, when not much was known of the terrible events inside the Soviet Union, many intellectuals admired the ideals of the Communist Party. Some of Oppenheimer's friends were members of the party, but he never joined it himself.

As he approached his fortieth year, Oppenheimer knew that he had not achieved any great distinction in physics. But then came his chance to work on one of the most dramatic applied science projects of all time. In the early years of the war, together with other U.S. nuclear physicists like E. O. Lawrence (1901–1958), the inventor of the cyclotron, Oppenheimer began to work on the theoretical possibility of an atomic bomb. In 1941, he calculated how much of the isotope uranium-235 would have to be brought together to produce an atomic explosion (the "critical mass"). By 1942, the U.S. government had at last put its weight behind the Manhattan Project, and General Leslie Groves of the U.S. Army was put in charge of it. Groves heard of Oppenheimer's research and was so impressed with his scientific and organizational abilities that he had him appointed as scientific director of the bomb project despite the objections of the security services, who were concerned about Oppenheimer's left-wing contacts in the 1930s. Oppenheimer exercised his persuasive charm to induce many of the best scientists in America to join him in this highly secret project at Los Alamos in the southwestern desert.

From 1943 to 1945, he helped overcome the many obstacles that lay in the way of a usable bomb: the theoretical problems; the technical difficulties; and last but not least, the management of a disparate group of hitherto purely academic scientists working under military discipline in uncomfortable conditions and cut off from communication with the outside world except under censorship. It was not surprising that relationships were sometimes strained. This was especially the case with Edward Teller, who had already foreseen the theoretical possibility of a fusion (hydrogen) bomb and seemed more keen to work on this idea than to get on with the job that was immediately at hand.

By April 1945, the "Alsos" mission of the U.S. Army advancing across Europe had established that the Germans were nowhere near producing an atomic bomb. Thus, the original motivation for producing the bomb—to save the world from Nazi tyranny—evaporated. One scientist, Joseph Rotblat, left the project at this stage, but the rest carried on. The political and military authorities were not going to abandon the investment they had made just when there seemed to be real hope of a very powerful new weapon. And the scientists were fueled by the technical interest—and indeed, excitement—of seeing what they had worked on so long and so diligently brought to fruition.

The first atomic test took place on July 16, 1945. The date was chosen in the hope of strengthening President Truman's bargaining position at the Potsdam conference. The Americans were considering using the bomb against Japan, and at this stage, Oppenheimer was coopted to an advisory scientific panel on the possible uses of the new weapon. As we have just seen, Szilard and others argued for a demonstration explosion followed by an invitation to surrender, but it seems that Oppenheimer did not object to the proposed use on Japan. The first bomb was dropped on Hiroshima on August 6, 1945, and the second (which was fueled by plutonium, not uranium) on Nagasaki a few days later.

Oppenheimer, like many of his colleagues in the Manhattan Project, began to feel a certain sense of guilt about the mass destruction they had made possible. In a lecture at MIT in 1947, he spoke, in one of the memorable phrases he was so good at coining, of the physicists "having known sin." Of course, the final decision to drop the bomb was made not by the scientists but by President Truman (with whom, as he said, the buck stopped). But the president had been given advice, and Oppenheimer had been a member of the advisory panel. A weapon constructed for one purpose—to prevent Nazi domination of the world—came to be used for another—to induce immediate and unconditional Japanese surrender. Many have argued that another strong motivation was to impress the Soviet Union, which was already seen as the Western powers' main rival in the postwar world.

After the war, Oppenheimer nominally returned to academic life but increasingly played a political role as an expert adviser behind the scenes. The newly founded Atomic Energy Commission (AEC) formed a General Advisory Committee (GAC) on scientific and technical matters, and he was the natural choice as its chairman. In the world of Washington politics, he began as an amateur but quickly picked up many of the rules of the game. In political infighting, however, his caustic wit and haughty dismissal of opposing views made him some powerful enemies. And meanwhile, in the atmosphere of "cold war" with the USSR, the FBI, in its persistent bureaucratic way, was maintaining its surveillance of Oppenheimer and building up its files on him.

The most important decision on which he next had some influence was whether the United States should make a hydrogen bomb, which would use nuclear fusion rather than fission and be much more powerful than the atom bombs used in Japan. Edward Teller had been enthusiastic about this idea ever since it was first conceived, and his chance to push for its development came when the USSR exploded its first atomic device in 1949, much sooner than expected. Teller and his supporters began to lobby members of the GAC, but Oppenheimer thought a "super" bomb would not be feasible, being too big to be deliverable by air and of no more military use than a large fission bomb.

Moreover, he was worried that the possibility of an even more powerful weapon would capture the imagination of Congress and the military as the only possible answer to the Russian threat. The GAC concurred with this opinion, and two of its members (Fermi and Isidor Rabi) went further, adding a minority report that condemned the H-bomb on ethical grounds. But this position was to hold for only two months. Pressure groups including Teller were working behind the scenes to reverse it. Then it was disclosed that Klaus Fuchs had been passing atomic secrets to the Russians for many years, including even Teller's ideas for the hydrogen bomb. Shortly thereafter, in early 1950, President Truman announced the decision to proceed with work on the H-bomb.

Teller went ahead with enthusiasm, of course, but the project soon faced technical difficulties that looked insuperable. Only when Teller came up with a new method for setting off the fusion of the H-bomb in 1951 were these overcome, and when this breakthrough was explained to Oppenheimer, he said (in another of his famous phrases) that the idea was now so "technically sweet" that going ahead with it was now irresistible; they would think about what to do with it after its success (Rouzé 1963, 83). This is hardly the language of someone with objections on moral principle to weapons that could destroy large cities at one blow. It seems as if Oppenheimer's earlier objections to the H-bomb were more pragmatic than ethical.

Oppenheimer's defeat over this policy question was followed by the notorious security hearing of 1954. As the Cold War with the USSR developed in the late 1940s and early 1950s, suspicion of anyone in the United States with even remotely left-wing tendencies grew to paranoiac extremes. Senator Joe McCarthy, in particular, led the movement to find "communists" and have them removed from their jobs. At this time, J. Edgar Hoover's FBI presented its files on Oppenheimer to President Eisenhower. They documented Oppenheimer's left-wing associations in the 1930s, his duplicity when interrogated by security officials in the 1940s (he had tried to protect friends and acquaintances from their inquiries), and his opposition to the H-bomb more recently. All this was used to argue that he should no longer be trusted with matters of national security. He elected not to resign voluntarily from his official positions but to face a full hearing, which soon took on all the worst aspects of a trial. His past was combed through with merciless attention to every unflattering detail. Many distinguished scientists testified on his behalf, but in one of the most crucial pieces of testimony, Edward Teller, while not accusing Oppenheimer of disloyalty, said he did not trust his judgment and would prefer that matters of national security were in other hands. The verdict went against him, and all his government connections were severed. He retained his academic position as director of the Institute for Advanced Study at Princeton University (where Einstein was), but despite his devoted service to the war effort, he was now publicly disgraced.

One obvious moral of this fascinating and many-faceted tale is that the scientists who develop an application of science cannot ensure that it will be used only for ends they approve. Another is that the scientist who tries to exercise political influence is as much subject to the vicissitudes of political fortune as anyone else.

Teller: H-Bombs and Star Wars

If one measures influence in terms of billions of dollars spent, Edward Teller (1908–   ) may well be the most powerful scientist of all time. He has been the prime scientific mover behind two enormous projects of U.S. defense policy—the creation of the hydrogen bomb in the 1950s and more recently, the Strategic Defense Initiative (SDI), or "Star Wars" as it was popularly known. It would seem to be not merely his scientific expertise that has given him this influence but the strength of his personality and his anti-Communist attitude, which won him much favor with U.S. politicians and military leaders.

Teller's political attitudes were presumably formed early in life as a result of his bitter experiences with anti-Semitism and dictatorship in Europe. Because he was Jewish, the brilliant young Teller was denied entry to universities in his native Hungary. He had to go to Germany for higher education (as well as his compatriots Leo Szilard and John von Neumann, who also ended up in the United States). Teller was one of the many visitors to Bohr's Institute of Theoretical Physics in Copenhagen, and he also formed a friendship with the aristocratic German physicist Carl F. von Weizsacker. But as Hitler rose to power, Jews were persecuted by students with Nazi sympathies, even in hitherto peaceful German universities like Goettingen. Teller found sanctuary in the United States. His experience of discrimination and persecution clearly left him with a strong feeling of revulsion against totalitarian regimes, which he later applied to the Soviet Union.

Like some other physicists, Teller was vividly aware of the possibilities of nuclear fission, and he and Szilard were the ones who tracked down Einstein at his Long Island vacation home in the summer of 1939 to get his signature on the famous letter to President Roosevelt. But Teller was soon thinking further ahead. Sometime in 1941 the idea of a much more powerful bomb occurred to him and Fermi over lunch: Perhaps a fission "trigger" could create enough heat to induce hydrogen atoms to fuse together, as they do in the sun, releasing vastly more energy than even a fission bomb, which was itself still only a theoretical possibility. The following year Teller joined a group to study the theory of atomic bombs, and after a while he began to monopolize the discussion with talk of the "super" bomb, at a time when they were nowhere near a practical fission version.

Then they thought of a still more frightening possibility. Calculations suggested that a fission reaction might produce enough heat to induce fusion between hydrogen and the nitrogen in the atmosphere. Could the first atomic bomb ignite the air itself and thus put an end to all life on earth? It would be an understatement to say that much depended on getting these esoteric theoretical arguments right! Compared with blowing up the whole world, even Nazi domination would be the lesser evil—after all, the latter could hardly last forever. Someone detected an apparent flaw in the calculations. But how sure could they be of either verdict? At one point, they estimated the chances of atmospheric ignition at three in a million, a risk that some of the group were apparently prepared to take. Teller claimed afterwards that his mistake in calculating that an atomic bomb might ignite the air was quickly corrected, and that the question did not worry them thereafter. (This story dramatically illustrates the problem of what attitude to take in the face of a small, not precisely quantifiable risk of a stupendous disaster—a problem that recurs in many other contexts, such as evaluating the safety of nuclear reactors, the release of genetically engineered organisms into the environment, and nuclear deterrence as a defense policy.)

Teller was summoned to join the Manhattan Project at Los Alamos in 1943. He admired Oppenheimer as its director but was unable to get along with him, and neither would he conform to the military discipline imposed on scientists working there. He seems to have always been a maverick, temperamentally incapable of taking orders from anyone. He was supposed to be helping produce the first fission bomb, but he was obsessed with the possibility of fusion. In the end, all the Los Alamos authorities could do with him was to let him get on with his own ideas independently of the main project. His relationship with Oppenheimer deteriorated. Even their memories later disagreed, for example, over their attitudes to the petition organized by Szilard asking the president not to use the bomb on Japan without giving them a chance of surrendering first—which Teller claims to have supported.

After the war, when Oppenheimer had left Los Alamos, Teller was offered the second most important post there. But he judged that he would not have been free to work on his beloved "super" bomb, so he went to the University of Chicago instead. From there, he campaigned for his pet project, arguing that just as Hitler could have gotten the atomic bomb first, so now there was a risk that the Soviets would beat the United States in the race to develop the hydrogen bomb. Teller found little support until the explosion of the first Soviet fission bomb in 1949, after which his persistent promotion of the H-bomb found the political wind blowing in its favor. There was a brief protest against the decision to make a hydrogen bomb by scientists as distinguished as Einstein, Szilard, and Hans Bethe, but to no avail. Such was the political atmosphere of the time that an issue of *Scientific American* containing arguments against making thermonuclear weapons was confiscated as dangerous to national security. The H-bomb has since been a fact of military and political life.

Ironically enough, Teller was not involved in the final birth of his "baby." Though he returned to Los Alamos to work on the development stage, he (once again!) did not get along with his colleagues there, and so he left. Thus, it came about that the prime inventor of the ultimate weapon sat alone on the coast of California to watch for the shock wave that was evidence of its first explosion far out in the Pacific. Shortly thereafter, Teller and his influential friends in Washington were pressing for the foundation of a second nuclear weapons laboratory. The proposal found favor with the air force (interservice rivalries were, and remain, an important part of the hidden infighting behind weapons policies). So started the Lawrence Livermore Laboratory, with Teller as its guiding light (although he formally occupied the post of director for only a brief period). At last he had his own show to run.

At the time of the notorious security hearing about Oppenheimer, Teller was the only scientist to testify against him. His most damaging statement was, "If it is a question of wisdom and judgment, as demonstrated by actions since 1945, then I would say one would be wiser not to grant clearance." Teller's long-standing resentment of Oppenheimer's influence over scientists and politicians—and in particular, of the obstacles he put in the way of the development of the H-bomb—were compressed into that one crucial sentence. What was supposed to be a decision about security clearance had become a political trial, a test of loyalty to what was now the government line. After the hearing, Teller was ostracized by much of the U.S. physics community for his part in the affair.

Through the years since those notorious episodes, Teller has remained at the forefront of U.S. military physics, ensconced in his base in the Lawrence Livermore Laboratory. He has been invited to deliver many prestigious lectures and has adapted some of them into books for the lay reader. In these works, he has argued for the necessity of nuclear deterrence and the safety of nuclear power and asserted the duty of the scientist to find new knowledge and explore its possible applications. He wrote that the only other duty of the scientist is to explain the knowledge and techniques as clearly as possible, leaving it to the people as a whole to decide what use, if any, to make of them (see the quotation from him in 2.6). This, of course, sounds admirably democratic—who could disagree with it? But it is worth asking how far this ideal was applied to the actual development of the H-bomb or of the atom bomb that preceded it. Were the American people given any choice? The crucial decisions to develop, test, and deploy the bombs were made before the public knew about the matter at all (see also 10.4). Influential individuals like Teller lobbied vigorously for certain decisions. He is not the sort of scientist who does his work and then sits back and hands it to the people (or their elected representatives), saying, "Here are the scientific results; you do with them as you think best." Obviously, he had very definite views about how science should be applied and maneuvered very effectively over the years to get his ideas accepted by those in power.

The same pattern is found in Teller's more recent brainchild, the SDI program, which President Reagan first announced to a wondering world in 1983. The prospect was put before the American people of an effective shield against incoming nuclear missiles that would shoot them down, à la *Star Wars*, by means of a computer-based system of nuclear-powered lasers. Who could criticize the ideal of defense, after all? Well, apparently many people. Criticisms have varied from straightforward doubt that any such system could ever be perfectly effective (for if one or two missiles get through, millions of people may still die) to suspicion that the whole thing was only a cynical public relations front for an attempt to defend U.S. missile sites so that a nuclear war could be "won."

Whatever the full truth of this matter, one thing is certain: SDI was not dreamed up by Ronald Reagan the day before his speech. Teller and his associates had been at work for many years, exploring the technical possibilities and trying to keep the military interested. Vast sums were expended, and the program was continued, though with less impetus, under President Bush. Many academic scientists refused to participate and campaigned against it. Most now agree that it never could have worked. In 1985, it was shown at Bell Labs to be theoretically impossible for SDI to function as a complete shield against incoming missiles, since the decision-problem for the computers would have been mathematically insoluble. Research then switched to much less ambitious partial defense. Some may argue that the whole thing was less sinister than it seems, being America's peculiar way of funding fundamental physics research and giving government subsidy to the economy. Defenders of the program now argue that it was a brilliant bluff, because U.S. expenditure on SDI drove the USSR into bankruptcy in trying to keep up with the arms race and so brought about its downfall. But that is wisdom after the event—SDI was obviously not originally conceived as a bluff.

The full story of SDI remains to be told, and the verdict is not yet clear. But it is not a case of the public deciding how to use science—and hardly of the elected representatives deciding, either. For better or for worse, certain scientists like Teller, if they make the right contacts and have politically favored attitudes, can sometimes exert a significant influence on government policy. They cannot then pretend to have no more responsibility than any other citizen.

9.2 Genetic Engineering and the Lobbying Scientist

Biology has come of age. It was once looked down upon as a junior among the sciences. The jibe earlier in this century was that except for physics, science was only "stamp collecting," that is, natural history that describes the contents of nature without theorizing about them (but that verdict hardly did justice to Darwinian evolutionary theory). But in its new, molecular form, biology has become one of the most swiftly advancing areas of science. After World War II, there was a discernible movement of scientists from physics into biology. This shift occurred partly because of the feeling expressed by Maurice Wilkins (whose part in the DNA story we mentioned in 5.1) that military applications had spoiled the beautiful subject of physics. But many sensed that biology was where some of the most interesting new scientific research and exploration could be done—witness the attitudes of Crick and Watson, also described in 5.1. In the 1950s and 1960s, following up on the elucidation of DNA, molecular biology was probably the fastest-growing area of science. Biology has now reached the stage where commercial (and even military) applications are burgeoning. The result is that biologists have had to face up to awkward questions about the financial, social, and political context of their work, as we shall now see.

Watson's Defense of Genetic Engineering

We followed the brilliant start of James Watson's career in 5.1. In the 1970s, well established as director of the prestigious Cold Spring Harbor Laboratory, he entered a phase of public controversy brought about by the advent of the first techniques of genetic engineering. In 1973, Stanley Cohen, working at Stanford University in collaboration with Herbert Boyer of the University of California, performed a pioneering series of experiments that suggested the feasibility of splicing genes from one organism into another. Having extracted the gene-carrying DNA from bacteria, they found that a restriction enzyme would cut the long DNA molecules in certain places, that a ligase enzyme would stick the pieces together again, and that it was then possible to insert the recombined DNA into other bacteria. In later variations on this theme, Cohen and Boyer spliced a section of DNA from a higher organism (a toad counted as "higher" for this purpose) into the DNA of a bacterium, and the toad gene was copied in new generations of the bacteria. Thus, bacteria, and perhaps higher organisms, might be reprogrammed and remodeled by the insertion of extraneous genes.

As soon as these techniques were shown to be practically effective, worries were expressed about the safety of using them on microorganisms, even under controlled conditions in the laboratory. Paul Berg and his group at Berkeley thought of splicing together part of a cancer virus with a bacterial virus that could get into the bacteria inhabiting the human intestine. Was there not a risk that extremely dangerous bacteria might inadvertently be created? Might they turn out to be carcinogenic or highly infectious and perhaps resistant to all known antibiotics? In one of the more lurid scare stories at the height of this public controversy, Mayor Vellucci of Boston frightened many people when he told some of his political enemies at Harvard: "God knows what's going to crawl out of the laboratory! They might even come up with a disease that can't be cured, or even a monster! Is this the answer to Frankenstein's dream?" (quoted in Langone 1978, 110) Overdramatized as such fears may have been, the molecular biologists themselves could not prove for certain that disastrous side effects would not result from applying the new techniques. But they had an obvious professional interest in going ahead, for they were eager to use newly available tools to do all sorts of scientifically interesting (and potentially rewarding) research.

This is a type of situation that has occurred elsewhere (for example, consider the debate over the safety of nuclear power stations), and it will recur again and again as science and technology proceed. The powers that be—whether they be in government, industry, or science itself—have an interest in using new technologies, and though they typically try to reassure the public by saying that in their expert opinion the risks are very small, they cannot (when intelligently pressed) prove that there is *no* danger at all. So who is to decide whether the risk is worth taking?

The debate over the dangers of recombinant DNA technology brought out different attitudes toward our relation to nature. At issue was the wisdom of such a very basic kind of intervention in biotic nature, with the prospect of guiding, as it were, the forces of evolution itself. Nobel laureate George Wald, in a public debate sponsored by the National Academy of Sciences, opposed recombinant DNA research in general, arguing: "A primary danger is that the deliberate production of 'desirable' changes in certain organisms might result in the disruption of the infinitely complex and delicate balance among living things, a balance which has slowly evolved in nature. A sudden disturbance of natural relationships could precipitate a disastrous and irreversible breakdown of our ecosystem" (1976, 10). Similarly, Robert Sinsheimer, then professor of biology at the California Institute of Technology, warned against introducing man-made organisms into "our precious and intricately woven biosphere" (1976, 2540).

The concern about recombinant DNA experiments was a special case of the general worry that interference with a complex natural system might disrupt delicately balanced relationships, with disastrous results. The commonly expressed warning is that we should not interfere with natural systems when we lack sufficient knowledge of the consequences. The response by some scientists that recombinant experiments were not really dangerous appealed to concepts of what is "natural." Berg and Singer argued: "There is no scientific fact or theory that predicts or supports the existence of a natural barrier (to gene interchange by different species); there is no reason to suppose that scaling this hypothetical barrier with recombinant DNA methodology would have the consequences Sinsheimer imagines" (1976, 2542).

The case of recombinant DNA research was an unusual one in that the scientists themselves proposed a pause in research while the implications of the new technique were assessed. In a letter to *Science* published on July 26, 1974, Berg and a number of leaders of the field, including Watson himself, proposed a moratorium on all experiments (of certain carefully specified kinds) using the new gene-splicing techniques until the possible risks could be evaluated more precisely. This restriction could only be voluntary, for although it came from the U.S. National Academy of Science's Committee on Recombinant DNA, it did not have the force of law in the United States, and there was no way it could have been enforced worldwide. Nevertheless, it was observed for a while, providing a rare example of scientists voluntarily holding back from the usual race for priority in discovery. The moratorium did have a certain moral force in that anyone who tried to jump the gun on his or her scientific rivals by experimenting in the supposedly dangerous areas could benefit from doing so only by publishing the results—but then it would be clear to the scientific community that he or she had broken the informal rule.

The Berg letter led to a famous conference about the safety issues in genetic engineering, attended by most of the world's leading molecular biologists, at the Pacific resort of Asilomar in early 1975. Understandably, they were far from unanimous in their views. Some vigorously claimed a right to free scientific inquiry, but others such as Sinsheimer continued to argue the need for safety restrictions. As Sinsheimer wrote afterwards:

> Here, I think, we have come to recognise that there are limits to the practice of any human activity. To impose any limit upon freedom of inquiry is especially bitter for the scientist whose life is one of inquiry; but science has become too potent. It is no longer enough to wave the flag of Galileo. ... Would we wish to claim the right of individual scientists to create novel self-perpetuating organisms likely to spread about the planet in an uncontrollable manner for better or for worse? I think not. (1975, 151)

The majority of attendees at the Asilomar conference, worried that more irksome legal restrictions might be put on their proposed research if they did not act to police themselves, agreed to classify certain experiments as riskier than others and to recommend two types of safety rules: physical (involving specific laboratory handling procedures) and biological (using strains of bacteria that were to be artificially enfeebled to reduce their chances of survival should they escape into the natural environment). The matter became one for public debate, political action, and bureaucratic procedures. Amid widespread suspicion of the whole governmental and scientific establishment (this was, after all, the era of the Watergate scandal, which forced the resignation of President Nixon), many of those most active in public campaigning were keen to put restrictions on research. The U.S. National Institutes of Health issued formal guidelines, restricting use of recombinant DNA techniques by drafting an elaborate set of rules.

Watson, along with many of his colleagues, soon came to feel that encouraging the public debate had been a mistake. "We were teenage jackasses," he said of the scientists (including himself) who had called for the moratorium. He meant that by crying wolf, they had created undue public concern over entirely hypothetical dangers. Watson argued that instituting rules restricting experiments to certain so-called safe areas gave a quite misleading impression that all other experiments were unsafe. The issue became deeply politicized, and Watson (never one to mince words) said that the whole episode had given far too much publicity and influence to "an odd coalition of spaced-out environmental kooks and leftists who see genetics as a tool for further enslaving the masses" (Watson and Tooze 1981, 346).

There ensued an effective lobbying campaign to relax the restrictions on recombinant DNA research, which was supported by scientists who wanted to go ahead with research and by would-be biotechnologists who foresaw profits in applying the new techniques (as we'll see in the next section). Watson played a vigorous part in persuading Congress not to impose further limitations on research and to relax those already imposed. As experiments and applications have proceeded, the initial fears have not been borne out, and Watson's judgment seems to be vindicated. Nothing terrible has happened—yet. Both the doomsday threats of unleashing epidemics and the promise of wonder cures for cancer and disease turned out to be greatly exaggerated. The worries expressed about genetic engineering in the 1990s tend to be more about ethical, legal, and commercial aspects of genetic engineering than about the safety of laboratory experiments. However, as biotechnology is scaled up to industrial levels and operated by less careful hands than those of research scientists, the potential for biohazards will certainly need watching. The deeper a scientific theory gets into nature, the more power it offers to human intervention, and the more possibility there is of something going wrong on a big scale. The disruptive effects of introducing foreign (not genetically engineered) species into different ecosystems remind us how easily environmental balances can be upset. The opposition of campaigners such as Rifkin, crude as they sometimes are (see 2.2), may work as a safety mechanism.

Cohen's Part in Lobbying Congress

Stanley Cohen was one of the first molecular biologists to demonstrate the feasibility of genetic engineering. He was also one of the earliest to realize its financial possibilities and apply for a patent. That there were tempting commercial prospects was already obvious. If human genes could be spliced into bacterial DNA, then perhaps the altered bacteria could be used as tiny biological factories—to make medically useful substances, for example. Maybe the genes of plants or farm animals could be reprogrammed to increase productivity.

It seems that Cohen was at first reluctant to apply for a patent and thus be open to charges of trying to make money from his scientific discoveries, especially at a time when the controversy over the safety of genetic engineering was reaching its height. But he was persuaded by the people at Stanford University's Office of Technology Licensing that if his gene-splicing process were *not* patented, the new techniques would more likely than not become trade secrets, hidden within private companies rather than open to public use through patents. Accordingly, the application was filed just before the deadline of a year that U.S. patent law allows after first public disclosure. The proposed patent was not to benefit Cohen and Boyer as individuals but the two universities of Stanford and California. Nevertheless, when this move became public knowledge in 1976, consternation arose that these two institutions were proposing to make a profit from a discovery whose basis had been laid by so many other people's work and funded largely by taxpayers' money. Lawsuits were threatened by those who had collaborated with Cohen and Boyer but were not included in the patent application. The patent was, however, issued in 1980 (under the Frankensteinian title "Process for Producing Biologically Functional Molecular Chimeras"—a chimera is a mythical monster with a lion's head, a goat's body, and a serpent's tail). It allows academic scientists to use the gene-splicing techniques without charge. And for commercial applications, the universities' lawyers cunningly set the license fees at such a level that it was more attractive for potential users to pay up than to try to evade the patent; as a result, the income from it has been substantial, at least by university standards.

These controversies over the safety risks and the commercialization of recombinant DNA techniques have been extensively documented, and we already have reviewed them. The episode we shall now consider occurred in 1977, when would-be genetic engineers like Cohen helped persuade the U.S. Congress to water down safety restrictions on their work. This is an instructive example of scientific expertise being used to exert political influence, in this case to further the interests of molecular biologists. (The courts face a growing problem of so-called junk science, where each side in a case calls in its purported experts and no effective procedures exist for deciding the reliability of supposedly expert testimony.) As we have seen, the moratorium on recombinant DNA experiments aroused public fears that many scientists dismissed as irrational. A second consequence was that proposed U.S. legislation inconvenienced the research that academic scientists wanted to do, while the safety of industrial processes using recombinant DNA was not considered. But rather than addressing these wider problems, many of the molecular biologists set out to persuade U.S. legislators that the proposed rules were unduly restrictive of scientific research. They had many motives—intellectual curiosity, certainly, but also an interest in furthering their scientific careers. And as we have noticed, financial interests were not a million miles away. (Like many others, Cohen had become associated with commerce by joining in 1975 the scientific advisory board of Cetus Corporation, now merged into Chiron.)

Now we come to the episode we promised to highlight. In 1977, draft bills were put before Congress proposing various degrees of regulation for experiments on DNA. In an attempt to persuade senators and representatives to adopt the least restrictive legislation, some new research findings were presented showing that the dangers of recombinant DNA had been much exaggerated. One of the papers most useful for this purpose was written by Cohen and his co-worker S. Chang; in it, they reported the spontaneous recombination, inside a living bacterial cell, of pieces of DNA from different species that had been experimentally inserted into the cell. But since the paper itself was as unintelligible to lay readers as most modern scientific publications are, it had to be summarized in ordinary language, and this is where the scientists' own interests appear to have influenced the result. Members of Congress and the press were given to understand that recombination of DNA can happen quite commonly in nature, that it must have been going on for ages, and that it therefore can hardly be dangerous. The *Washington Post* (September 28, 1977) quoted Cohen as saying that "Mother Nature has been capable all along of doing in cells what scientists can now do," and senators were heard repeating this line. But the Cohen-Chang experiment had not proved that DNA recombines randomly in nature all the time; it had only shown that if bits of DNA were artificially introduced into a cell, they *might* combine there. No direct evidence existed that fragments of DNA could get through cell walls in nature. However, even those like René Dubos, who said that he initially had "a quasi-religious hostility to experiments combining genes from different organisms, because I felt that this was contrary to the ways of nature," came to realize that "genetic exchange occurs frequently in natural conditions" and that it is likely that bacteria "can incorporate DNA fragments from the animals and plants in which they reside" (1977, 238).

Cohen was accused by some of misrepresenting the situation to serve his interests. Dr. Roy Curtiss III, professor of microbiology at the University of Alabama (who was among the most cautious of the scientists at the beginning of the safety debate, though he has later revised his views), went so far as to describe Cohen's part in this affair as "one of the most imperious, despicable, pieces of political science that I know of" (quoted in Lear 1978, 242). Whatever the misinformation was, the general concern raised by this incident is that the molecular biologists interpreted scientific results in such a way as to influence politicians to do what the former wanted, namely, to pursue their research without legal restrictions. The scientists were no longer disinterested seekers after truth; they had become players in the games of power, in science, politics and business. Cavalieri (1981) labeled the whole episode "From Truth to Power."

The Human Genome Project

A vast new phase in the story of molecular biology has opened with the Human Genome Project. Surprisingly, this ambitious plan for mapping out all human genes was first proposed by a group of physicists at the nuclear weapons lab at Los Alamos, who presumably were concerned about the future of their jobs and proposed switching from studying radiation damage to doing something more positive. In 1985, Sinsheimer, who had become chancellor of the University of California at Santa Cruz, was looking around for some large project to help put his university on the scientific map. He must have overcome his earlier trepidations about genetic engineering, for he invited some of the world's leading experts in molecular genetics to discuss the idea of building a new institute for mapping and sequencing the approximately three billion "letters" of the human genetic code. Many of the scientists thought this impracticable, but (as we have seen in 7.2) Walter Gilbert was sufficiently impressed by the project to propose undertaking it with his private Genome Corporation.

Congress began to take an interest in the Human Genome Project, both as a grand scientific-medical initiative that could be described as "reading the Book of Human Life" and as something that could help maintain the American lead in biotechnology for years to come. In 1988, Congress allocated $3 billion to be disbursed over a period of fifteen years to complete the project. James Watson found himself at the very center of the issue, and once again, by accepting the position of director of the Office of Human Genome Activities within the U.S. National Institutes of Health, he demonstrated his commitment to scientific research, apparently unfettered by social worries. His recent resignation from this position amid allegations of conflicts of interest involving shares in biotechnology companies provides further evidence of just how ethically and politically fraught the whole thing has become.

Controversy continues about whether this project represents the wisest use of scientific funds at this time, and there are concerns about the possible uses of such knowledge of human genetics. Sinsheimer had already expressed trepidation about the rapidity of scientific advance: "How far will we want to develop genetic engineering? Do we want to assume the basic responsibility for life on this planet—to develop new living forms for our own purpose? Shall we take into our own hands our own future evolution?" (1975, 151).

Knowledge gained from the Human Genome Project promises (or threatens) to give us the power to tinker with the basis of human nature itself. Attitudes toward such knowledge exhibit strong ambivalence; on one side, the conservative fear of unknown dangers counsels restraint, but curiosity, the drive for adventure, and the scientific and technological urge to try all possibilities bid some people to press on. It is difficult for a scientist to conceive that certain matters are best left unknown, at least for a time.

Watson, as one might expect, is enthusiastic, remarking in his contribution to Kevles and Hood's *The Code of Codes* (1992) that "I don't want to miss out on learning how life works" (165). Interviewed on the British television science program *Horizon*, Watson said (with the "eloquence" that the medium tends to induce): "It's too interesting. To get the answer. It's like arguing for motherhood. I mean, you want to know it. Because it—the DNA—is, you know, the program for human existence. So, really to understand ourselves, we're going to really have to understand our DNA." Along similar lines, *The Code of Codes* also contains an article by Gilbert entitled "A Vision of the Holy Grail." Although the phrase "of the Holy Grail" was inserted by the editors (but not, it must be admitted, rejected by Gilbert—see Mueller-Hill 1993, 406), some find such hyperbolic claims to be rather disturbing. In what sense can learning more about the molecular processes of genetics help us really understand ourselves, not to mention the meanings and values in human existence? Critics of the project argue that even if full sequencing of the human genome is achieved—at considerable cost—the real work lies in the interpretation. The development from DNA in the germ cells to the whole organism is not as simple as following a blueprint. The relation between the genes and the characteristics they "code for" is complex, generally involving interactions among a number of different genes and environmental factors.

Besides technical difficulties in achieving the promised therapeutic results, a number of ethical issues have been raised. Some observers are worried that research and applications will be overinfluenced by commercial interests. As we have seen, the issue of conflict of interest was a factor in Watson's resignation in 1992 as head of the NIH Office of Human Genome Activities. Others are deeply disturbed by the possibilities of human beings acquiring control over their own genetics (see our mention of Rifkin's views in 2.2). Hans Jonas viewed the project as a threat to "the integrity of the human image." Fearing entry into unknown territory with unpredictable consequences, he pleaded that scientists resist the "Promethean urge to open Pandora's box" (1985, 337). In a review of nine works on the Human Genome Project, Harvard population biologist Richard Lewontin expressed deep skepticism about how the project will be used. He sees the "ideology of biological determinism" lurking in the kind of enthusiasm for the project expressed by Watson and Gilbert (1991, 72).

The project is now being sold to the public by the promise of a revolution in medicine, where knowledge of the structure and function of genes would lead to the control of genetic diseases and genetic predispositions to cancer and other maladies. Watson and Cook-Deegan stated that "the principal goal of the human genome project is to assist biomedical researchers in their assault on disease" (1990, 3322). When genetic "engineering" is construed as "gene therapy," it sounds much less threatening—as suggested by H. Vasken Aposhian (1970, 106). Consideration of just one case is impressive. A single, recently located gene is said to cause breast cancer in 85 percent of women having it. Because 10 percent of all women are estimated to carry the gene, about half a million women in the United States alone could possibly benefit from an early warning. Gene therapy has already been attempted. In 1990, a four-year-old child suffering from a rare hereditary blood disease was provided with an enzyme missing from his immune system by injecting genetic material by means of a virus vector. But laymen (and -women) are understandably confused when scientists disagree with each other about the wisdom of pursuing the Human Genome Project. The public's ideas of the benefits and dangers is often based on grossly inadequate conceptions of the ways that genetics works. In a 1986 Harris poll, almost two-thirds of the U.S. population surveyed confessed that they knew very little or nothing about biotechnology. Nevertheless, 44 percent approved the project of altering the genetics of children in the future to make them more intelligent or better at sports. For all its scientific sophistication and promised benefits, the Human Genome Project looks like it is making some already difficult social and medical issues even more complicated.

9.3 Ecology and Crusading Scientists

Ecology is the study of interrelations between organisms and their environments. The "new ecology" (systems ecology) emphasizes functional relations among the basic units of an ecosystem, which consists of a large number of interdependent populations of organisms living together in an environment. This branch of science has only recently become prominent, though it has roots in the work of Darwin. It examines how the organisms in an environment interact to make up a large and complex whole, as opposed to physiology and molecular biology, which study the subparts and processes within individual organisms.

In ecological studies, it is often hard to draw a clear line between objective, disinterested descriptions of a species or an ecosystem and expressions of concern for its preservation and functioning. We all have ideas about what makes a human individual (or society) stable and healthy, and it is well-nigh impossible to keep such value judgments out of biology when we study the functioning of plants and animals. Ecologists and ethologists have typically displayed a tendency to campaign for the conservation of their subject matter: ecosystems and their constituent species. The concern may be simply to preserve the environment to maintain its economic value for future generations. Many environmentalists today, however, promote the idea of "deep ecology": that natural systems should be regarded not just as resources for human use but as having intrinsic value, independent of their relation to us. For example, British television ecologist David Bellamy (formerly a lecturer in botany at Durham University, until the media discovered his popularizing talents) was to be found risking arrest for obstructing the construction of a dam that threatened a unique ecosystem in Tasmania, and American ethologist Dian Fossey became so protective of the mountain gorillas whose behavior she was studying in their natural habitat in Africa that she alienated the local population and was murdered.

Professional ecologists and ethologists are not the only scientists to show concern for the environment and to get involved in public campaigning. We have seen in 8.2 how Russian physicist Andrei Sakharov, worried about the radioactivity produced by H-bomb testing in the atmosphere, bravely began to press his political masters to desist from it, which led him into wider criticisms of the socioeconomic system of the USSR. We shall now look at some examples of such environmental crusading by scientists in the West.

Carson: The Web of Life

More than anyone else, Rachel Carson (1907–1964) can be described as the mother of the environmental movement that began in the 1960s. Her book *Silent Spring* (1962) awakened public concern about the hazards of the indiscriminate use of pesticides and thereby aroused general interest in environmental issues.

Carson received a master's degree in marine biology from Johns Hopkins University in 1932 with a thesis on the embryology of the kidney in catfish. She wanted to do scientific research, but as a woman she faced the usual difficulties of the time in getting a decent position in science, whether at a university, in private industry, or with the government. She did manage to get a position with the U.S. Bureau of Fisheries (later the Fish and Wildlife Service), and her talent as a writer led to her becoming the bureau's editor-in-chief of information service. She began writing on the side, partly to help support herself and her mother. Carson became something of a literary celebrity when her book *The Sea Around Us*—with its beautiful descriptions of the ocean's ecosystems—was serialized in *The New Yorker* in 1951. This was followed in 1955 by *The Edge of the Sea*, which made vivid the intricate interconnections that support the balance of nature. Carson's writings are excellent examples of combining the best of the literary and scientific cultures.

Her passionate protest against the unconstrained, destructive use of pesticides in *Silent Spring*, serialized in *The New Yorker* in 1962, raised a storm of controversy. Carson was moved to write the book by reports of the devastating effect of DDT on bird populations. DDT had been used effectively in World War II to eradicate the mosquitoes that caused malaria and in agriculture to control a variety of insect pests. There was hope that DDT would be the miracle chemical that would rid the world of harmful insects. The U.S. Department of Agriculture enthusiastically promoted its use, along with various other poisons that were by-products of wartime research on chemical weapons. Only a few observers saw danger in the unrestrained use of such chemicals. In 1945, Carson's suggestion to *Reader's Digest* of an article on the effects of DDT on the environment was rejected. She pursued her inquiry, however, spending four years gathering evidence of the damage that pesticides caused to the natural environment. Her theme was the conviction that "in nature nothing exists alone" (Carson 1962, 51). Her strong rhetoric against the indiscriminate use of pesticides is illustrated by her comment on an incident of spraying beetles in Illinois that resulted in the poisoning of a number of mammals and birds; such cases "raise a question that is not only scientific but moral. The question is whether any civilization can wage relentless war on life without destroying itself, and without losing the right to be called civilized" (99). She depicted the uncontrolled use of pesticides as warfare against nature, rejecting the traditional notion that nature exists only for the convenience of humans.

The agricultural chemical industry reacted vehemently to what it perceived as a threat to its existence. Carson was accused of being nothing more than a hysterical woman, a sentimental nature lover without professional credentials who was trying to wreck the agricultural economy. On CBS reports in 1963, spokesmen for the industries that manufactured pesticides condemned *Silent Spring* as being guilty of "gross distortions of the actual facts, completely unsupported by scientific experimental evidence and general practical experience in the field" (McCay 1993, 81). But Carson's specific arguments were never refuted, even if her rhetoric might be criticized as exaggerating some of the possible consequences of the indiscriminate use of pesticides. She never opposed all uses of them but was attacking the lack of sensitivity to their complex and sometimes very destructive effects on ecosystems.

Public pressure motivated by *Silent Spring* led in 1963 to a report on pesticides by President Kennedy's Science Advisory Committee, which supported most of her arguments. Carson's beautiful and persuasive writing aroused public interest in environmental concerns that previously were largely confined to a relatively small number of conservationists. Her sounding the alarm helped bring about legislation to protect the environment, in particular the establishment of the Environmental Protection Agency (EPA).

Environmental Crusading by Commoner and Ehrlich

Barry Commoner (1917–) and Paul Ehrlich (1932–) are two American biologists who have devoted much of their professional work to drawing public attention to the problems of pollution and overpopulation. Commoner was one of the first scientists to become a public campaigner not just for any particular ecosystem but for the ecosphere as a whole (though he carefully documented particular cases such as the death of Lake Erie). He was trained as a biochemist and started his scientific career in an orthodox manner, doing research on the chemical and physical basis of processes characteristic of functional living cells and on the biosynthesis of tobacco mosaic virus. But when he became director of the Center for the Biology of Natural Systems at Washington University, St. Louis, he used his position as a platform for what might be described as a crusade to change the world—or at least to alert everyone's attention to the need to change worldwide social and economic practice. Commoner argued that the scientist as "custodian of the knowledge everyone needs to exercise his own conscience" has the responsibility to "get the information to the public" (1971b, 181).

In a series of books written for the general public, including *Science and Survival* (1966) and *The Closing Circle: Nature, Man and Technology* (1971a), he documented the increasing threat to human well-being and survival caused by our systematic misuse of the ecosphere. He sees pollution of the environment and the overpopulation problem in the Third World as twin symptoms of a worldwide system of economic exploitation.

There seems to be a (perhaps suspiciously) smooth transition here from scientific analysis of what is actually happening in the worldwide ecosystem to a radical politicoeconomic diagnosis of what needs to be changed. Commoner argued that the "present course of human civilization is suicidal" (1971a, 295), and he could see only two options for us: (1) "rational social organization of use and distribution of the earth's resources" (requiring central government control) or (2) "a new barbarism" (298). His concern is to alert everyone to the problem rather than to recruit his readers into any particular political party. But naturally enough, he has been involved in social, legal, and political action. He founded the St. Louis Committee for Atomic Information, which tried to ensure that the public's voice was heard in inquiries into the siting of nuclear power stations. Later this group became the Committee for Environmental Information, which published the magazine *Environment* and helped show that nerve gas that the U.S. Army proposed to dump into the Atlantic could be safely detoxified instead. Commoner has also served as chairman of the board of directors of the Scientists' Institute for Public Information and was involved in the successful legal action in 1971 to compel the U.S. Atomic Energy Commission to disclose information about the environmental effects of its proposed fast breeder reactors.

The line of argument that Commoner has followed is indicated by the metaphor of "the closing circle." We have, he contended, "broken out of the circle of life" (12). He enunciated "laws of ecology" in an effort to shift our attitude toward nature from the Baconian view, in which it is something to be conquered and mastered by humans for our own uses, to a respect for nature, with human beings holding "stewardship" of the earth. Commoner's first law of ecology, "Everything is connected with everything else," expresses his holistic view of nature (33). The suggestion is that if we disturb natural systems, there is a danger that they will collapse. Aldo Leopold (1887–1948), an early leader in American conservation, expressed an even more far-reaching, holistic view of values in nature in a famous chapter entitled "The Land Ethic" in his book *A Sand County Almanac*: "A thing is right when it tends to preserve the integrity, stability, and beauty of the biotic community. It is wrong when it tends otherwise" (1949, 262).

Public environmental debates are generally not conducted in scientific discourse. The appeal is rather to the feelings by means of symbols that have emotional appeal. We seem to need such symbols to motivate practical action. For example, in the highly publicized debate over whether the possible extinction of the spotted owl in the U.S. Northwest would be worse than the loss of jobs if logging its habitat were curtailed, the owl is seen by those wishing to preserve the habitat as a symbol of general environmental degradation, the destruction of wildlife habitats, and decreasing biodiversity.

The holistic vision held by many environmentalists sometimes inclines them to take an all-or-nothing view of policies. For example, in arguing for a ban on all logging in federally owned virgin forests in the United States, along with banning all use of herbicides, pesticides, and fertilizers, Tim Hermach, head of the Native Forest Council, used this analogy: "You can't save a cancer patient's life by incrementally removing the malignant cancers from his body. You either go in and take out 100 percent of the malignancies or the patient dies" (Letto 1992, 26). A general problem in informing the public about environmental problems is striking a balance between effectively arousing interest—which may involve dramatizing the dangers by sketching scary scenarios of consequences if we fail to take action—and accuracy in conveying the best knowledge, indicating the possible dangers and outlining policies with some indication of relative risks, costs, and benefits.

If the only motivation for preserving the environment is an appeal to self-interest, the problem of the "tragedy of the commons" (Hardin 1968) may arise. If access to a publicly owned resource—like a common pasture—is unrestricted, people may be tempted to overuse it, thus destroying the resource (as a pasture may be ruined by overgrazing). Paradoxically, the very attempt by individuals to maximize their benefit (as "rational" egoists) leads to a loss to everyone. Hardin saw this as a pervasive problem wherever costs are commonized but benefits are privatized. For example, general pollution of the environment may result when "the rational man finds that his share of the cost of the wastes he discharges into the commons is less than the cost of purifying his wastes before releasing them" (1968, 1245). Hardin's solution is "mutual coercion mutually agreed upon," say, by imposing taxes on pollution that would motivate limiting its production.

Paul Ehrlich, professor of biology at Stanford University, made a scientific reputation by his work on butterflies, in which he examined their morphology and classification as well as their reproductive biology and its relationship to larval foodplants. But he rapidly extended such academic study of population biology to concern about our future on this planet and became a public crusader like Commoner. Ehrlich made a specialty of documenting the human population problem in books such as *The Population Bomb* (1968) and *The Population Explosion* (1990), written jointly with his wife Anne. He founded an action group called Zero Population Growth, and in his heyday he was giving as many as a hundred public lectures a year and many more radio and television broadcasts. He had a tendency to overstate his case deliberately by painting doomsday scenarios of famine and war to gain attention and arouse concern. Some of his forecasts have been far off the mark, having more the character of prophecies than scientifically based predictions. The danger of such crying wolf, of course, is that the crier loses credibility and the importance of his message may be missed.

Ehrlich saw the so-called green revolution as a paradigmatic case of a "technological fix" for social problems that ultimately makes things worse. Geneticists succeeded in breeding new strains of the major food grains, wheat and rice, which promised to help solve the problem of feeding the world's growing population. Cultivation of these new grains was successful in greatly increasing their yield in Third World countries. But many problems arose from unintended consequences, illustrating the maxim that you "can't do just one thing." These more efficient grains required more chemical fertilizer, which is not only more energy intensive but can also result in pollution problems, with runoff into rivers and lakes affecting their biota. In the process of selecting for efficient production, genetic diversity was decreased, with the consequence that the new strains tended to be less resistant to insects and disease. So their use results in greater use of chemical pesticides with potential dangers to humans and the ecosystem. Economically, the green revolution tended to serve the interests of agribusiness, leading to the formation of large estates that made the rich richer and the poor poorer. Those who owned all this land found that it was more profitable to raise cash crops for export than for domestic consumption, leaving less food for the native population. Finally, populations have continued to increase, creating the need for yet more efficient food production. The results were mixed: Food production was increased but not without unintended side effects.

Should scientists themselves pay more attention to solving social problems? The popular conception of scientists as wizards with magic formulas might suggest that we simply let them study our problems and come up with solutions. To plan consciously, we need knowledge. But complex systems, from individual organisms to ecosystems, pose great, perhaps insuperable, difficulties for accurate scientific analysis and prediction. The record of scientists (and nonscientists) in predicting the consequences of new technologies is not very encouraging. In his 1955 presidential address to the American Association for the Advancement of Science, Warren Weaver cited some of the reasons why living nature does not yield to the kind of analysis that has been so impressive in the physical sciences: "Physical nature . . . seems on the whole 'loosely coupled,' that is, simple physical systems can be analyzed into parts, focusing on two or three isolated variables at a time, and we often can use linear approximations to get useful information." Organisms, by contrast, are highly complex coupled systems, so "it takes many variables to describe a man or a virus" (1955, 1258).

Many scientists have gone on ecological campaigns since the pioneering days of Commoner and Ehrlich. The worst scenarios have not yet materialized, but the problems of pollution and population are obviously very far from solved. Some changes in public awareness, legislation, and industrial practice have been achieved (not counting merely cosmetic changes like calling an arm of government "the Department of the Environment," or introducing "green" themes in advertisements on TV). Presumably, some credit for this must go to these initial crusading ecologists—though it is very hard to be sure about causal connections between social phenomena.

Toward the end of his incisive book *Scientific Knowledge and Its Social Problems* (1971), Jerry Ravetz hailed Commoner's work as a model of what he calls "critical science" as opposed to "industrialized science," the now more typical form of scientific research funded and directed by the great powers of state or industry (see 1.4). The "critical" scientists are those who are not controlled by such powerful paymasters. They have the financial and political independence to be able to document in a scientifically professional manner various facts (for example about pollution or industrial or military safety hazards) that may be appropriate matters for public concern but that may be very inconvenient for the powers that be to acknowledge or have publicized. And since there is little point in doing this work unless the results are made public, critical science will of course involve an element of politics. The hope is that action will follow naturally from analysis.

A recent example is Jeremy Leggett. While a geologist at London University, he devoted some of his time to exposing the fallacies in official arguments that underground nuclear tests were not reliably detectable (at the time, the Ministry of Defence was trying to justify the British government's opposition to the Comprehensive Test Ban Treaty). Leggett has since left academic employment to become chief scientist of the campaigning organization Friends of the Earth. The era of professional critical science has arrived.

Suggestions for Further Reading

9.1

Blumberg, A., and G. Owens. 1976. *Energy and Conflict: The Life and Times of Edward Teller.*

Easlea, B. 1983. *Fathering the Unthinkable: Masculinity, Scientists and the Nuclear Arms Race.*

Jungk, R. 1958. *Brighter Than a Thousand Suns: A Personal History of the Atomic Scientists.*

Michelmore, P. 1969. *The Swift Years: The Robert Oppenheimer Story.*

Pais, A. 1991. *Niel Bohr's Times in Physics, Philosophy and Polity.*

9.2

Lear, J. 1978. *Recombinant DNA, The Untold Story.*

Cavalieri, L. F. 1981. *The Double-Edged Helix* (especially ch. 7).

Kevles, D. J., and L. Hood, eds. 1992. *The Code of Codes: Science and Social Issues in the Human Genome Project.*

Krimsky, S. 1982. *Genetic Alchemy: The Social History of the Recombinant DNA Controversy.*

Lowrance, W. W. 1986. *Modern Science and Human Values* (ch. 5).

Watson, J. D., and J. Tooze. 1981. *The DNA Story: A Documentary History of Gene Cloning.*

Yoxen, E. 1983. *The Gene Business.*

9.3

Carson, R. 1962. *Silent Spring.*

Commoner, B. 1967. *Science and Survival.*

————. 1971. *The Closing Circle: Nature, Man and Technology.*

Dixon, B. 1973. *What Is Science For?*

Ehrlich, P. 1968. *The Population Bomb.*

Ehrlich, P., and A. Ehrlich. 1970. *Population, Resources, Environment: Issues in Human Ecology.*

McCay, M. A. 1993. *Rachel Carson.*

Merchant, C. 1992. *Radical Ecology: The Search for a Living World.*

Ravetz, J. 1971. *Scientific Knowledge and Its Social Problems.*

Tobias, M., ed. 1985. *Deep Ecology.*

10

Science and Values

Our discussion of various contrasting perceptions of science in Chapter 2 reflects the remarkable ambivalence people have shown toward science; it can be the subject or source of unbounded optimism or profound pessimism, hope or fear, adulation or repugnance. Sometimes science seems like a bountiful goddess and sometimes more like a demon. For some thinkers, it is an impersonal, remorseless force that undercuts all values, yet for others it represents an influence promoting distinctively humanist values.

The common human tendency is to think in black-or-white terms, especially where our values and emotions are involved. The subtle questions about science and human values surely deserve a more discriminating response. Yet most people find it hard to cope with a complex, nuanced, undogmatic, provisional, open-ended answer. Perhaps none of us can entirely escape the tendency to polarize our feelings, attitudes, and beliefs—but we can make an effort to enrich our understanding of some of the difficult questions about science and its uses and misuses. This book is designed as an encouragement, and a tool, to do just that. In this concluding chapter, we shall take a more critical look at those various images of science and scientists that we surveyed in Chapter 2.

10.1 Optimism, Pessimism, or Critical Realism?

It should be obvious by now that we cannot rest content either with the Baconian optimism toward science discussed in 2.1 or the Frankensteinian pessimism presented in 2.2. In their simple forms, both attitudes focus on one part of the truth but ignore other, crucial aspects of the total picture. The application of science has already produced manifold human benefits, that much is clear—and the potential exists for many more. But science has also been put to many negative, destructive ends, most obviously in warfare and weaponry but also in the unintended, primarily secondary effects of radioactivity and many other kinds of pollution. Yet even in the seemingly negative cases, the overall cost-benefit result is far from obvious; if some argue that warfare would have been much more destructive since 1945 if we had *not* had the deterrent effect of nuclear weapons or that our standards of living would have been much lower if we had *not* had the products of those industries that have also polluted our environment, can we prove them wrong?

We will have to learn to live not only with the uncomfortable fact that there can be both beneficial and harmful applications of science but also with the even more uncomfortable thought that it is often difficult to tell which are which. However, there is no need to sink into a skeptical or hopelessly quietist attitude that what will be, will be, and there is nothing we can do about it. It is perfectly possible, and surely most humanly rational, to maintain an open, yet critical attitude:

1. We should welcome the benefits that science and technology can offer, while being aware that not all so-called benefits are as unproblematic as they seem and that even the best-planned technologies have unforeseen, unintended, and unwanted consequences;

2. We should also be aware that there will always be forces in the world, both in individuals and in institutions, that will tend to use the powers of science for their own ends, whatever those may be, and not necessarily for the benefit of humanity.

The verdict that any particular technological development is or is not worthwhile must therefore be forever provisional and open to reevaluation in light of new evidence. In this way, no permanent answer can be given to questions about the uses of science, any more than about the truth of scientific theories themselves.

When we consider the motivations of scientists, we had better be prepared to recognize that these, too, will often be mixed and complicated, not open to evaluation in simplistic black-or-white terms. The pattern displayed in each of our case studies in the central chapters of this book illustrates this point. In all of them, we can find more than one motive operating. Has the vision faded of the scientist as a dedicated, incorruptible researcher, persevering in the quest for truth without thought of reward? We have found the ideal of pure curiosity almost completely realized in some of the classic scientists such as Mendel, Marie Curie, and Einstein, who were discussed in Chapter 4. But scientists are only human. It would be unrealistic not to recognize other, more worldly motives that they exhibit: the need to earn a living; competition for appointments, promotion, grants, and prizes; ambition to influence the development of their subject; the temptation to make big money by patenting inventions or setting up commercial companies; and the attractions of status, fame, influence, and power in the wider international community. We have seen all these tendencies in Chapters 5 through 9.

A few scientists may exhibit the obsessive, aggressive, destructive, macho tendencies expressed in the Frankensteinian image—just as some people (mostly men?) do in the wider community. But there seems to be no good reason to believe in any general association of such personality traits with scientists. If anything, the reverse would seem to be the case. The motivation of pure curiosity, the strong desire for knowledge just for its own sake, and a consequent respect both for the evidence of observation and for the rational arguments of other scientists are still prominent, especially (even necessarily) among the most creative scientists. Of course, scientists have always exhibited a variety of motivations. But some observers are concerned that with the industrialization of science (see 1.4), the aims of science are shifting away from these traditional ideals because of the way scientific research is currently conducted and funded. We shall return to this topic in 10.7.

10.2 Does Science Really Undercut All Values?

What about the charge (raised in 2.3) that science undercuts all values? Many have voiced this accusation. Counterculture historian Theodore Roszak elaborated Blake's entreaty "May God us keep/From Single vision & Newton's sleep!" arguing that the scientific vision presents each of us with "the experience of being a cosmic absurdity, a creature obtruded into the universe without purpose, continuity, or kinship" (Roszak 1972, 154). Philosopher Hans Jonas sees the "category of the sacred" as "most thoroughly destroyed by scientific enlightenment" (1984, 23). In a recent best-seller, British science journalist Bryan Appleyard called for a "humbling of science" because it has humbled man by robbing him of "the myths, metaphysics and illusions necessary for art" (1992, 170). Appleyard complained that science has created "the vision of man as a fragile, cornered animal in a valueless mechanism" (112) and that this "hollow mechanistic vision . . . now threatens to destroy us" (248). A diametrically opposite view was expressed by American physicist Isidor Rabi in a lecture at Columbia University on "The Philosophical and Social Implications of 20th Century Physics"; according to Rabi, science is "the only valid underlying knowledge that gives guidance to the whole human adventure. Those who are not acquainted with science do not possess the basic human values that are necessary in our time" (quoted in *Time*, May 26, 1967).

Of course, much more could be said about these large issues than we can attempt in a book like this. Philosophers have debated the nature of values for more than two millenia—and there is today an explosion of publications on the subject. But when beginning to reflect on these matters, we should realize that if we agree that in some sense scientific theories give us truths about the world, we need not accept the view that there are *no* different, equally valid perspectives on reality. Such "scientism," a perjorative label for the view that knowledge obtainable by scientific method exhausts all knowledge, goes far beyond science. It is not an empirically justified scientific theory but a controversial philosophical thesis to posit that whatever is not mentioned in the theories of science does not exist or has only a subordinate, secondary kind of reality.

Making the distinction between primary and secondary qualities (explained in 2.3) need not involve us in saying that the latter do not exist. Colors may not be mentioned in theories of physics, or if they are, they are explained in terms of differential reflection of various wavelengths of light. But that does not show that is it untrue, wrong, or misleading for us to continue to describe things as colored. The same goes for the other secondary qualities: sounds, heat and cold, tastes and smells. To explain why *X* exists in terms of what in the physical structure of the world makes *X* occur is not to explain *X* away, as if *X* were not really there to be explained after all.

Science explains an event by exhibiting general connections between some of its properties and the properties of other events. In thus *relating* a given property to others, scientific explanation may give us a new understanding of the property, but this does not eliminate its reality. Much the same argument can apply to many other qualities that more obviously involve value. Take aesthetic values, for instance. We describe certain combinations of colors or sounds as harmonious or discordant, exciting or drab, happy or sad, conventional or original. Are these value-laden terms necessarily undercut by the thought that what we thus describe can also be described in terms of physical properties of wavelengths and amplitudes of light or sound? Accepting the truth of the physical story does not imply that there is no justification for aesthetic responses and evaluations.

Creative scientists often find inspiration and great beauty in what others see as dreary or threatening. The fact that the scientific pursuit of knowledge should proceed without personal bias does not imply that a scientist should lack emotional interest in her work. The physicist Richard Feynman saw great value in the scientific worldview that can lead one "to imagine all sorts of things infinitely more marvelous than the imaginings of poets and dreamers of the past" (1958, 262). He felt inspired by the vastness of space and time, the dance of molecules, and the evolution of life. For Feynman, science has the thrill of a "grand adventure"; the awe and mystery of nature is for him deepened by scientific knowledge. Noting that few unscientific people seem to share his feelings about science, he lamented that "this is not yet a scientific age" because poets and artists rarely try to deal with the world of science (262). The reason why more people aren't singing science's praises, he surmised, is that they can't read the score of the music. Most people are interested in the practical results of science, not in the ideas behind the applications.

What many critics find lacking in science is "ultimate explanations," that is, explanations in terms of purposes. The Italian humanist scholar Giambattista Vico (1668–1744) claimed in *The New Science* that we cannot really know the external world through science; we only know what we are acquainted with "from the inside," such as our own motives and purposes. Thus, we can know human affairs in a way that we cannot know nature. According to Vico, only God can "know" nature since He created it, and only the one who makes a thing truly understands it (1982, 198). Thus, true knowledge of art and literature is direct, immediate perception of particular things in individual experience. A similar argument was put forward by poet Archibald MacLeish, who tried to specify a kind of knowledge that only literature can provide. He firmly rejected Bertrand Russell's statement of scientism: "Whatever can be known can be known by means of science" (1945, 834). MacLeish insisted that "poetry is also capable of knowledge . . . a kind of knowledge of which science is not capable." He saw science as giving us knowledge "by abstraction." Poetry does not abstract; it rather "presents the thing as the thing," which gives us (in English poet Matthew Arnold's words) "a full, new and intimate sense of it" (MacLeish 1956, 49–50).

A desire to reject scientific reductionism can already be found in Socrates' reaction to the speculative natural philosophers of ancient Greece. When asked if he believed a certain legend, Socrates, in Plato's *Phaedrus*, replied:

> I might proceed to give a scientific account. . . . If our skeptic, with his somewhat crude science, means to reduce every one of the legends to the standard of probability, he'll need a deal of time for it. I myself have certainly no time for the business. . . . I can't as yet 'know myself,' as the inscription at Delphi enjoins, and so long as that ignorance remains it seems to me ridiculous to inquire into extraneous matters. (1961, 229c)

In the Platonic dialogue, *Phaedo*, Socrates stated:

> I once heard someone reading from a book by Anaxagoras, and asserting that it is mind that produces order and is the cause of everything. This

explanation . . . made me suppose, to my delight, that in Anaxagoras I had
found an authority on causation who was after my own heart. . . . As I read
on I discovered that the fellow made no use of mind . . . but adduced causes
like air and aether and water and many other absurdities. It seemed to me
that he was just about as inconsistent as if someone were to say, the cause
of everything that Socrates does is mind—and then, in trying to account
for my several actions, said first that the reason why I am lying here now
is that my body is composed of bones and sinews, and that the bones are
rigid . . . and since the bones move freely to their joints the sinews by re-
laxing and contracting enable me somehow to bend my limbs, and that is
the cause of my sitting here in a bent position. (1961, 97b)

Socrates' concern was to emphasize that some reasons for human actions are not matters of mere physiological causation. He abandoned
physical investigations in favor of philosophical inquiry into ethics and
politics—seeking, for example, the essential nature of justice. (Philosopher of science Karl Popper [1963] saw Western philosophy taking a
wrong turn here and advised a return to the pre-Socratic concern with
the natural world.) The Socratic hope was expressed in the dictum "virtue is knowledge," though not the kind of knowledge offered by natural
science.

What then are we to make of moral values and social values in the
light of science? Suppose we accept that in some sense it may be possible in principle to describe and explain the workings of human bodies
and brains—presumably in the vocabulary of physiology and cellular
biology and perhaps of genetics and molecular biology as well. Does this
invalidate the evaluation of some attitudes and actions as better than
others, as more worthy of admiration, emulation, and promotion? Once
again, the possibility of one level of description need not undercut the
rationality or intelligibility of another. In practice, not even the most
hard-nosed exponent of what is sometimes called "the scientific attitude" can avoid making evaluations of his own and other people's behavior in terms of rationality, ethics, or political wisdom. Scientists
themselves certainly evelute each other in these ways, like everybody
else, as we have seen in many of our case studies.

10.3 Science and Technology: Humanizing—or Dehumanizing?

Some traditional humanists scoff at Snow's claim, outlined in 2.5, that
industrialization is the hope of the poor, that scientific technology has
generally bettered the conditions of the masses and is the only way of
making further progress in this direction. Although we may deplore the

very real evils of the sweatshops and child labor of the early Industrial Revolution, we should also note that people came from the country seeking such urban employment opportunities as the lesser evil. We can get a glimpse from Daniel Defoe's journals of 1725 of the often brutal conditions under which the majority of men, women, and children lived before the late eighteenth century in England. And we should remember how average standards of nutrition, health, and longevity have risen substantially over the past two centuries in the industrialized West. We might ask whether contemporary problems caused by pollution are worse for the average person than the health risks that accompanied the state of plumbing and garbage disposal in medieval towns? Those who express nostalgia for life in a prescientific culture tend to assume that they would be members of the privileged elite, not laborers or serfs.

On examination, it is not very clear what a cry of "back to traditional humanist values" really means. Has the understanding offered by traditional culture outside science improved the lot of the average person? We must note that some of the greatest atrocities in history have been carried out not by those committed to scientific values but by those purporting to be faithful to traditional religions: the Aztec mass sacrifices, as well as the massacres of the Aztecs themselves and other Native Americans by the conquistadors; the Inquisition burning heretics at the stake; and the massacres of the Armenians in Turkey in 1915. In recent times, large-scale abominations have been committed in the name of political ideologies that hardly reflected the influence of science: the starving of millions of Ukrainians in Stalin's Soviet Union, the inconceivable horrors of the Nazi death camps. Cruelties and slaughters of prescientific history are so numerous as hardly to attract attention. On the other hand, news bulletins regularly remind us that the science and technology of our own time have not prevented the worst of human nature from manifesting itself in horrible ways.

Critic Lionel Trilling found "bewildering" Snow's claim that the traditional literary culture largely "manages the Western world" (1962, 465). Does this make the traditional, nonscientific culture "answerable for all the anomalies, stupidities and crimes of the Western world?" he asked (468). Snow didn't really say this, though he did suggest that the traditional literary culture has contributed little to impede the stupidities and crimes of those in power. Another literary critic, F. R. Leavis, ridiculed Snow's "message, the sum of his wisdom" that there is "social hope":

What is the "social hope" that transcends, cancels or makes indifferent the inescapable tragic condition of each individual? Where, if not in individuals, is what is hoped for . . . to be located? . . . [D.H. Lawrence] diagnoses [Snow's confusion] in his supreme novel, *Women in Love* . . . he insists on the truth that only in living individuals is life there, and individual lives cannot be aggregated or equated or dealt with quantitatively in any way. (1962, 38–39).

Though Snow's way of putting it leaves him open to Leavis's criticism, we can surely make some sense of "social hope" for the individual, beyond her limited life span, in a feeling of participation in the prospects for the future generations or at least for her own progeny.

What then about the dehumanized anti-utopias mentioned in 2.4? It is arguable that such repressive dystopias do not truly represent the spirit of a scientific society. As we have seen in 3.1, science depends essentially on the attitude of critical questioning being applied to theories and beliefs about the world—involving a willingness to test them against the evidence of observation and to develop, amend, and even radically change them if necessary. Science must remain open to such intellectual progress; otherwise, prevailing theory becomes merely an orthodox system of belief, and science proper ceases to exist. One might argue that even the maintenance of some sort of social stability is in the long run more effectively achieved by the apparently less stable democracies, which allow changes both in governments and in ideas. Regimes that try to repress all political or intellectual change may succeed for a while, but when change eventually comes, it is likely to prove violent, uncontrollable, and revolutionary.

10.4 Is Science Value-Neutral?

As we saw in 2.6, one standard response to the question of how science relates to values is to say that it is fundamentally and necessarily value-neutral. A distinction is made between the theories about laws of nature that *pure* science gives us and *applied* science or technology, in which human beings manipulate and change the world, trying to fulfill their various desires or goals. Theories are thought to be a matter of knowledge, but values, which people try to promote by practical action, are often assumed to be a purely subjective matter and not a topic for knowledge. We distinguished three distinct theses within this belief in science's value-neutrality. Let us now consider them critically, in reverse order.

The applications of science are, to use an oft-repeated phrase, "for society to decide." But how can decisions about the proper ways to apply scientific knowledge be made by society? Can scientific research and its technological applications be under the democratic control of the citizens who ultimately pay for it and are affected by it?

There is really no such agent as "society" to make such decisions; that is, society is not an entity that can arrive at a decision, at least not when no practical possibility exists for all citizens to meet together and consider what should be done. The decisions of society are really the decisions of various institutions—governments, courts, corporations, banks, universities, churches, political parties, pressure groups, and so on—and, of course, the decisions of individuals. In democracies, there are processes of election—and in some countries, referenda on specific issues—by which a decision of some sort is extracted from a mass of individually expressed preferences. But this kind of mechanism, valuable and, indeed, essential as it may be for choosing a government and making constitutional changes, cannot for obvious practical reasons be used to produce a decision on every detailed question of the application of science.

There are good reasons for wondering how far contemporary scientific research and its technological application is (or could ever be) under the democratic control of the citizens. Much vital research is now conducted under conditions of military or industrial secrecy; by the time the results of such research become publicly known, it is too late for anyone to argue that the effort and resources might have been better directed elsewhere. Such can be the situation facing a newly elected politician being briefed about the technological programs that have been going on behind the scenes. "Experts" will advise him, "This is a project on which much has already been spent, which is soon to come to fruition, and which it would be madness to abandon at this late stage, just when we are about to gain an advantage over our rivals." The industrialized nature of so much modern scientific research and technology requires a long lead time for development, a large number of people with specialized expertise, and as a consequence, a great deal of money. As a result, a scientific project typically acquires a momentum of its own that any outside force, even a clear majority of public opinion, will encounter extreme difficulty in stopping.

Another factor favoring those responsible for directing research is that they can usually determine how the matter is presented to the public, because of its very technical nature and the secrecy that usually surrounds it. With the aid of those skilled in the ways of the mass media, public opinion can be "molded." Consider, for example, how the Strategic Defense Initiative (see 9.1) was presented to the American public as "Star Wars," thus associating it in the public mind with movies in which the goodies beat the baddies by clever technology. Who could disagree with SDI under that description?

Let us consider now the second point of the value-neutral conception of science: the claim that the only thing valued by the scientist is knowledge for its own sake. She would have to be a very "pure" scientist indeed who was content to hide her light under a bushel, who did not care in the least about her scientific reputation and professional advancement, not to mention the influence and rewards that success can bring. Very few of the scientists we have discussed would count as pure in this sense! But the areas in which funding, appointments, promotion, fame, and reward are to be found are determined by social forces outside the control of the individual scientist. In the past, perhaps, the deciding factor was simply the professional judgment of his peers, and the resources necessary for research might be provided by the average university laboratory. But in many areas of science, the picture has now changed enormously—the era of "big science" has come, as we noted in 1.4. The leading edges of research and development have got to the point where forward movement requires large teams and expensive equipment. This means that hard choices have now to be faced about the direction and funding of research. Because of the huge costs involved, the concentration of research into larger units and its control by large institutions seem inevitable. Since governments have become increasingly involved in this process, a political element now enters into decision-making even about the purest scientific research. Science is becoming a business; medical science, in particular, is now often spoken of as part of the medical industry. In his Herbert Spencer lecture of 1973, Popper expressed his concern over the change in the spirit of science that occurs when "too many dollars [chase] too few ideas." "Big science," he warned, "may destroy great science" (1975, 96).

So even though scientists may wish to say that their only professional commitment is to increasing human knowledge, they will now have to recognize that the funds for their research will probably be given with a close eye to possible applications, be they military, industrial, medical, or whatever. Doing such research under these conditions cannot be said to be value-free. By accepting funds from certain sources—and agreeing to make their results available to those funding them—scientists are participating in social processes by which knowledge, and hence power, is given to certain social groups rather than others, for example, to industrial corporations, defense departments, or national institutes of health. They may even have to make a difficult choice between doing their research under these conditions or not doing it at all. If they participate in the process as actually institutionalized, they display tacit acceptance of those institutions' values.

The Frankenstein image never was very plausible for the average scientist (and the rare fanatical individual is fairly easily controlled). What we surely need to worry about much more is the power of the institutions that increasingly direct scientific research and its applications: the research councils, the commercial companies, the rich private foundations, the armed services, and the government departments. Such bodies may be made up of reasonably well-meaning individuals, each of them earning their living and doing their duty as they conceive of it, yet the institutions can act like corporate Frankensteins, pursuing power or profit regardless of social consequences. The fear is that deeper scientific knowledge will tend to put more power into the hands of those who are already powerful and who may well misuse it. Governments, the military, and industrial corporations may acquire even greater capacity to affect (for better or for worse) our way of life, our food, our health, and the environment.

Finally, let us consider again the first element in the conventional picture of the value-neutrality of science: that science can only deal with objective facts, not values. A sharp distinction between facts and values has been commonplace in twentieth-century thought, not just in the philosophies of positivism and existentialism, which have dramatized it most, but as a background assumption that conditions much everyday thinking. This division raises a deep philosophical issue—and a highly controversial one—that we have already touched on in 10.2. The widespread assumption that all moral (and political) values are subjective should certainly not be allowed to pass without question; such a view represents a major claim in the theory of meaning, knowledge, and metaphysics—that an unbridgeable chasm lies between the standards governing scientific claims and those governing moral claims.

We shall not try to settle here the debates about the objectivity of values that have been going on at least since the time of Socrates. But it is worth noting that this thesis of the unique objectivity of science might be attacked in *two* different ways. It may be suggested that science itself does not really have the kind of objectivity commonly attributed to it, or it may be claimed that discussion of values can in principle be as objective as scientific discourse is commonly thought to be.

Examples of the latter kind of argument can be found in the work of Hilary Putnam and of Jurgen Habermas. Both writers question what Habermas (1971) called "scientism"—the positivist thesis that our very standard of what is to count as knowledge should be defined in terms of the natural sciences; they thus attack the first element in the conventional arguments discussed in the previous paragraph. Habermas recommended "reflection" on the ends of our actions and in particular on the applications that we may consider making of scientific knowledge. His hope seems to be that if the conditions of communication of knowledge, opinion, and argument are ideal, then the discussion of values can approach the standard of rationality commonly recognized in the sciences. We will next look at a proposed way of settling value questions rationally: cost-benefit analysis.

10.5 Numerate and Ecolate Thinking

In addition to changing the way people live by helping create new technologies, science has affected how people think, how they make decisions, and how social policy is made. The kind of economic cost-benefit analysis that is commonly applied today by government agencies in policy decisions evolved from attempts to apply scientific reason to social policy. Jeremy Bentham (1748–1832), an English writer and legal reformer, promoted what came to be called a *utilitarian* view of right decision. He proposed a principle of utility that bids us "approve or disapprove every action according to the tendency which it appears to have to increase or diminish the happiness of whoever's interest is in question" (1962, 2). His idea was that a course of action is right if it leads to the "greatest happiness for the greatest number" (142). Bentham viewed people as motivated by the "springs of human action": pursuit of pleasure and avoidance of pain (1). The benevolent legislator should therefore formulate policy with the aim of maximizing the general welfare as a sum of the interests of the individuals in the community.

In contemporary decision theory, a "rational" choice is commonly taken to be that which maximizes the agent's utility, given her preferences and beliefs, or in the case of social policy decisions, the course of action that maximizes the utility for society. If risk is involved, if the good or bad consequences are estimated to occur with various probabilities, then such factors are also taken into account. The principle of maximizing utility bids us choose that course of action with the greatest utility, where utility is calculated as the sum of the products of the probabilities and assigned quantitative values of the consequences of the action. In this approach to making decisions, science is supposed to contribute information concerning the effects of alternative courses of action along with their probabilities.

Once quantitative values are assigned to the various consequences, the calculation can be automatically made, yielding a "rational" decision. The procedure is as follows:

1. Identify a problem and formulate goals.
2. Specify a set of alternative courses of action.
3. Assign probabilities (risk factors) to the outcomes.
4. Assign values (numerical measures of utility) to each outcome in a consistent, systematic way.
5. Calculate the utility of each course of action: the net cost-benefit as the sum of the products of probabilities and values of the outcomes.
6. Choose the course of action with the greatest utility.

Applied to policy decisions, this procedure is supposed to yield potential benefits to society that outweigh the potential costs.

Such a "scientific" approach to making decisions has a number of rational virtues. It makes us more aware of the value judgments implicit in the choices; it encourages us to consider a variety of alternatives and their consequences; and finally, it takes risks into consideration in clear, quantitative terms. The rational decision advocate asks us to be numerate: Don't ask simply whether something is safe; ask *how* safe—what risks are socially acceptable? Once we raise this question, the Delaney Amendment to the Pure Food and Drug Act appears irrational—"scientifically indefensible," as Hardin (1985, 42) remarked, since it requires a total ban on any additive to human food that is shown to cause cancer in *any* fraction of *any* rodent at *any* concentration. We always face risks, so we should try to make the best trade-offs. The "objective" risk depends on the probability and severity of the consequences. Thus, the rational fear of flying should generally be less than the fear of driving on the highway. By 1950, the risk of death from smallpox in the United States was less than that from the vaccination against it, so it was no longer "rational" to get vaccinated.

The institutionalization of science has encouraged incorporation of risk-cost-benefit analysis into policymaking. Such analyses are required by law in the regulation of risks by governmental agencies such as the EPA and are commonly used today to evaluate alternative medical procedures. Risk-cost-benefit analysis has also been used by the courts to adjudicate government regulation. In 1989, the EPA banned the use of asbestos in various products such as piping and roofing. A federal court struck down the ban, arguing that it would cost millions of dollars while saving less than an average of one life per year. Judge Stephen Breyer noted that toothpicks, which cause about one death a year due to accidental swallowing, pose a greater risk than the asbestos products in question.

The "objective rationality" of cost-benefit analysis does not appeal to everyone. Norman Cousins expressed the revulsion felt by some literary humanists: "The world will end neither with a bang nor a whimper [as in T. S. Eliot's poem "The Hollow Men"] but with strident cries of "cost-benefit ratio" by little men with no poetry in their souls. Their measuring sticks will have been meaningless because they are not big enough to be applied to the things that really count" (1979, 8).

Biologist Lewis Thomas voiced a more moderate but still concerned reaction to the increasing use of cost-benefit calculations in environmental policy, noting that "it goes somehow against the grain to learn that cost-benefit analyses can be done neatly on lakes, meadows, nesting gannets, even whole oceans. It is hard enough to confront the environmental options ahead, and the hard choices, but even harder when the price tags are so visible" (1973, 121).

Serious questions can be posed about how cost-benefit studies are made, what they really tell us, and what their proper role in decision making is. What range of consequences needs to be considered, and how far into the future? How do we make allowance for future values? Do we have the right to impose our value judgments on future generations? There are problems with the very idea of quantifying certain values, for example, in putting a price on human life or trying to assess the worth of preventing a species from going extinct. We have intuitions—on which philosopher Immanuel Kant constructed his ethics—that moral human agents have a "dignity beyond price":

> In the realm of ends everything has either a *price* or a *dignity*. Whatever has a price can be replaced by something else as its equivalent; on the other hand, whatever is above all price, and therefore admits of no equivalent, has a dignity. That which is related to general human inclination and needs has a *market price*. . . . But that which constitutes the condition under which alone something can be an end in itself does not have mere relative worth, i.e., a price, but an intrinsic worth, i.e., *dignity*. (1959, 534–535)

Hans Jonas expressed Kant's view when he insisted that "only awe of the sacred with its unqualified veto" (1984, 23) goes beyond the calculations of utilitarian ethics to unconditional categorical judgments of what is right and wrong.

We often have to make decisions in situations of uncertainty, in which no reliable way can be found to assess the probabilities of some of the consequences because of unpredictable factors that may be involved, including human error and malice. This type of dilemma (for example, over the safety of nuclear power or genetic engineering) will recur again and again as science and technology proceed. The powers that be—whether government, industry, or the scientists themselves—may have an interest in using a new technology, and though they will typically try to reassure the public by saying that in their expert opinion the risks are very small, it is unlikely that they can demonstrate that there is no chance of danger at all. If there is a small risk of some utterly horrific disaster, who is to decide whether such a risk is acceptable?

In quantifying values, the hope is that they can be cashed out arithmetically in some common currency, but who is to assign those numbers, and what if people disagree about them? A risk-cost-benefit analysis cannot settle such questions without controversy. They are necessarily a matter of judgment, and different people will often make very different judgments. So either a solution is imposed by those who have the power to do so, or the issue becomes a matter of public politics. Questions of justice may arise concerning the distribution of costs and benefits. A strong ethical argument can be made for saying that it is those who actually bear the risks—those adversely affected if something should go wrong —who should decide whether the risks are worth taking. (This is part of the rationale behind informed consent in medical testing.) When the public at large is affected, the issue has to be political; it cannot be solved simply by a mathematical calculation.

Cost-benefit analyses are only decisive when the parties involved know the risks and can reach consensus on the values. Nevertheless, it seems eminently rational when making policy decisions to ask, What are the consequences and how do we evaluate them? Ecologist Garrett Hardin (1985) has suggested that in making policy decisions, we need both the "numeracy" of economic cost-benefit analysis and "ecolacy," which looks at the big picture and takes a long-range view of the consequences, recognizing human fallibility and the limitations of our knowledge.

10.6 Is Science Objective?

Let us now look at some arguments against science having the kind of objectivity that most people think it has. Objectivity has traditionally been considered to be a central virtue of science. After all, science is supposed to be public knowledge. Its findings are testable by any qualified observer. The methodology of science is supposed to safeguard against any bias of subjective feelings and emotions that might influence its judgments. But some critics see science as just one ideology among many. Paul Feyerabend (1978) has adopted a radically relativist position, according to which science is one "tradition" among many others, such as ancient or primitive worldviews and various religious or political belief systems. His crucial claim is that these various traditions cannot be rationally compared for truth, since all such judgments about truth or rationality can only be made from within one tradition. Along with this relativism concerning truth, Feyerabend has recommended "political relativism," namely, a "free society" in which all traditions are given equal rights and equal access to education and other positions of power. He sees this as involving a strict separation between science and state, like that presently acknowledged between religion and state in most Western democracies. He grants that early modern science "*was* an instrument of liberation and enlightenment," but he believes that science today "inhibits freedom of thought" (1975, 157–158). Part of his concern is the particular character of contemporary scientific research: "Most scientists today are devoid of ideas, full of fear, intent on producing some paltry results that they can add to the flood of inane papers that now constitutes 'scientific progress' in many areas" (165).

This splendidly provocative challenge to conventional wisdom about the supreme rationality of scientific method deserves a more careful answer than we have space to give here, but the following point is worth making immediately. If Feyerabend is to be entitled to make a distinction between comparing rival theories within a tradition and to allege the impossibility of any rational comparison between traditions themselves, he had better have a clearly articulated notion of identity for these so-called traditions. He has to be able to tell us when a change in concepts or beliefs is merely a change *within* a tradition and when it constitutes a change from one tradition to another. Which description, for example, would he apply to the Copernican revolution in astronomy, to the advent of Darwinian theory or of relativity-based physics, or to the difference between organic and psychodynamic accounts of mental illness? Unless Feyerabend can give us a principled way of answering such questions, his claims about the limitation of rationality within traditions have no clear content.

Those who, like Feyerabend, reject the very concept of objective truth can be accused of committing the slippery slope fallacy. From the fact that no one, scientist or otherwise, can be completely objective in the sense of achieving theory-free observation or value-free interpretation, it does not follow that the concept of objective truth is empty. Some of those who criticize science as biased—for example, in favor of patriarchal values—may hope to achieve a truly objective science by making us conscious of tacit value assumptions that distort scientific judgments. But if a proper science, free from prejudices and distorting values, can be conceived, we must first have some basic notion of objectivity that is violated by the bias or distortion.

Some years earlier, Herbert Marcuse, whose thought became briefly fashionable in radical movements of the 1960s, made a perhaps even more radical onslaught on scientific and technical rationality and its social consequences. He claimed that the way in which "scientific-technical rationality and manipulation are welded together into new forms of social control" is not just the result of a specific social application of science but was already "inherent in pure science" (1964, 146). He added that "the science of nature develops under the technological *a priori* which projects nature as potential instrumentality, stuff of control and organization" (153) and that "science, *by virtue of its own method* and concepts, has projected and promoted a universe in which the domination of nature has remained linked to the domination of man" (166). Yet he went on to suggest that science could somehow become radically different: "Its hypotheses, without losing their rational character, would develop in an essentially different experimental context (that of a pacified world); consequently, science would arrive at essentially different concepts of nature and establish essentially different facts" (166–167).

But Marcuse's vision of an alternative form of science, which would establish different concepts and facts from those presently acknowledged, also seems very schematic and philosophically undefended. He owes us an account of what he sees as definitive of science in whatever forms it may take and of which features of present-day science he thinks could be altered, and how. It is not clear (from his jargon-infested prose) that he does more than gesture at this. Of course, research might be pursued in some areas rather than others—for reasons of finance, social need, military pressure, ethical inhibitions, or scientific fashionability. But Marcuse's thesis appears to be (if we take seriously the last sentence quoted above) that even on a given topic, an alternative way of doing science would yield different theories about the nature of the world. And presumably he does not mean simply that there can be complementary theories of the same phenomenon (such as wave or particle theories of light). It seems that such differences would for him be within the "domineering" way of doing science, to which he wants to suggest a radical alternative. But unless he can characterize this in clearer terms than abstractions like "being," "logos," and "eros," we may wonder whether he really has any alternative to offer.

Nicholas Maxwell has more recently argued for a total transformation of our attitude to scientific knowledge. He wrote that instead of the "philosophy of knowledge" (which would seem to be his version of the conventional, value-neutral image of science that we have been questioning), we should espouse the "philosophy of wisdom," that is, we should "give absolute intellectual priority to our life and its problems" (1984, 65). Maxwell's idea thus seems to be that all our research should be aimed directly at goals of human value. But later it emerges that he is committed to more than a mere change in the *motivation* of scientific research, for he apparently wishes to dispute much of the widely accepted *epistemology* of science, which he labeled "minimal standard empiricism" (200). He claimed to show that "empirical considerations alone cannot decide what theories are to be accepted and rejected in science: metaphysical considerations concerning the comprehensibility of the universe must be taken into account in addition" (201). He thus raises issues of the basic epistemology of science, about which all we can realistically do here is recommend further reading.

We mention these various writers not just because they sense problems in the way that science is currently being applied but because they make interesting and controversial claims about how the roots of those problems lie in philosophical assumptions about the nature of science and of knowledge in general. These radical thinkers raise questions about science and its applications less commonly mentioned by most academically prominent philosophers of science. But it is one thing to ask good questions and another thing to give good answers to them. The brief comments we've just quoted already give us reason to wonder whether either Feyerabend or Marcuse has a coherent position to defend. The programmatic remarks of Rifkin (see 2.2) and Maxwell's "philosophy of wisdom" leave it unclear whether they want to question just the wisdom of our present applications of science or also the validity of scientific methods as the way to find out truths about the material world and how far the two kinds of criticism are supposed to be connected.

10.7 Values in Doing Science

In 10.4 we have questioned the conventional wisdom that science is value-neutral. Even if a scientist is motivated solely to attain knowledge for its own sake, some value judgments have to be made simply to engage in research. When research programs are chosen, the selection reflects a decision about what is worth knowing and thus reveals implicit values. Scientists don't collect just any facts at random. The average weight of the grains of sand on a beach or the size of the sandbugs might be estimated, but unless such facts are thought to be relevant to testing some wider theory of geology or biology, why should anyone want to bother wasting his time making intrinsically boring measurements? Even on topics where theoretical or practical interest is more plausible, like the patterns of clouds or the positions of stars in the sky or the bone structure of fossil fish, it is reasonable to ask how much time and effort is worth spending on such matters.

Why should anyone want to know something? For what reasons, to what ends, is a particular research project undertaken? These questions clearly involve values. Nobody expends time, effort, or resources without having some goal that is seen as valuable. All intentional human action—scientific activity included—involves goals or desires of some sort. So why should anyone want to learn about the inner workings of nature? People are sometimes curious about things for no apparent reason. Children typically ask lots of "Why?" questions, and adults may also wonder why the moon goes through phases, why the tide returns twice a day, or why a tortoiseshell cat can have kittens of quite different colors. If someone is "just curious," the motivation for wanting that knowledge is simply to satisfy a harmless human whim. But scientific research is motivated by more than idle uninformed curiosity, though it can start from that. Discoveries are sought for their theoretical interest or their practical applications—or both. We can thus distinguish three kinds of reasons for a scientist wanting to know something:

1. *Simple curiosity,* which for no further reason may set off inquiry

2. *Theoretical interest,* based on relevance of a problem to understanding and explaining other phenomena

3. *Potential usefulness* for achieving some practical human purpose

These are the obvious kinds of reasons scientists propose in favor of pursuing a program. But there may also be reasons *against* proceeding with certain research. About any scientific investigation we can also ask, What are the costs of finding out? Various *kinds* of cost may be involved in gaining an item of knowledge, not all of them monetary:

1. Any scientific inquiry, however humble, takes someone's time and effort. Even a nineteenth-century rural dean, botanizing on weekdays, could reasonably have been asked why he devoted himself to his pastime rather than ministering to the needs of his parishioners. Charles Darwin's family, though secure in his private income, may well have wondered why he had to devote so much time to his fossils and papers instead of spending more time with them (see 5.1 and 7.1).

2. Whenever research needs more than trivial resources, questions can obviously be asked about whether they might not be better used elsewhere, whether within or outside science. Lavoisier was not given the directorship of the national arsenal in eighteenth-century France just to

indulge his scientific curiosity (though he took full advantage of the opportunities thereby provided). The French government obviously expected him to be useful for military purposes (see 7.1). The German bankers and industrialists who gave money to establish the Kaiser Wilhelm Gesellschaft in 1911 made a judgment that this investment would prove useful to them and to Germany (see 6.2). Much modern research demands very considerable resources of labor, technology, energy, and therefore money. We are now at the stage where it is a matter of national and international controversy whether we should devote large resources to push back the frontiers of particle physics by building yet more powerful accelerators.

3. Sometimes there are human costs to the process of scientific research itself. Gathering data in dangerous places or doing experiments with radiation, poisons, viruses, or microbes exposes scientific investigators to obvious risks. There may also be risks to the wider public, for example, if some pest or poison or infection were to escape into the environment. A major problem about rationally evaluating such risks is that they often cannot be realistically estimated in advance of the results of the research itself. Consider, for example, the controversies in the 1970s about the new techniques of genetic engineering, which we considered in 9.2.

4. Sometimes research may violate basic ethical principles. Experiments on people can be conceived whose results might be scientifically interesting but that would be totally unethical to perform. For example, depriving children of various features of normal upbringing and social stimulation might help us discover to what degree human behavior is inborn, genetically determined as opposed to culturally variable. But any such deliberate deprivation of children is morally ruled out.

We may think that only Nazi doctors would do such things, but the recent revelations of experiments on the effects of radiation performed on members of the American public without consent through the 1940s and 1950s should make us aware of the constant temptation to break the bounds of ethics in the name of science.

In other cases, there may be disagreement, both among scientists and the general public, about whether the intrinsic interest and possible benefits of the scientific research should be allowed to override ethical principles in particular cases. Examples might be research on human embryos, experiments involving animal suffering and death, the issues of fairness to human patients involved in clinical trials of new drugs or medical procedures, issues of deceiving subjects of experiments in social psychology, and violations of privacy regarding the release of personal data about individuals.

Even when questions about the immediate costs and benefits and morality of a research project are answered, further worries may arise about how the knowledge will be used. The proposed *process* of research itself may be acceptable, but a different set of considerations may be applied to the consequences of its success. What are the likely uses and misuses of new knowledge?

1. In some cases, the intended uses may be beneficial, and yet there may be unintended disadvantages as well. A new drug may be designed to cure or alleviate certain diseases, but things could go wrong in many ways. It may have unpleasant side effects, which cannot always be foreseen. If a patent on a drug is about to expire, it may be in the company's financial interest to produce and patent a new version, regardless of whether the modification is of any medical benefit. It can also happen that the widespread use of individually effective medicines can have demographic effects beyond anyone's expectation. Should the fact that reducing infant mortality may contribute to overpopulation, with consequent starvation, lead us to restrict efforts to save the babies?

2. A scientist may want to know something, both for its own intrinsic interest and for the possibilities of beneficial applications, and yet the institutional or social situation may be such that she may have serious worries about likely misuse of the new knowledge. We can expect that government or industry will use new discoveries and inventions for their own purposes, with which the individual scientist may well disagree. (Physicists like Szilard first struggled to bring the possibility of an atomic bomb to the attention of the American authorities, only to find a few years later that they could not influence the military use of the new weapon—see 9.1). There may sometimes be good reason to predict that if a certain new technique is made publicly available in a society, it will be used in ways about which one might have serious moral qualms. An addictive tranquilizer that is of benefit in managing very disturbed individuals may be indiscriminately prescribed by overburdened doctors to speed the exit of unhappy patients from their offices. The availability of procedures to determine the sex of fetuses could lead to widespread abortion of females fetuses in countries where boys are preferred.

3. Even if we do not have *specific* misuses in mind, there may be social choices, which on the whole we might prefer not to have to face, that

could be forced on us simply by the further advance of scientific understanding. For example, however scientifically interesting the mechanisms of human genetics may be, do we really want to be given the opportunity of, and hence the responsibility for, deciding the genetic characteristics of our offspring? Again, would we welcome the technical possibility for employers and insurance companies to classify people according to their genetically based tendencies, as shown by a printout of their particular set of genes? It will be said that this is a matter for "society" to decide. *Who* in society? Do we want the necessity for such decisions to be thrust upon us? Do we want our political parties and legislatures to be bogged down with having to frame complex policies and laws for a host of such possibilities, when there are many other urgent matters for decision and social action?

So the process of scientific research cannot be value-neutral. The general reason for this is that like any other human activity, it involves choices of how to spend time, energy, and resources. The special reasons are peculiar to the high costs, institutional control, and social applicability of scientific research that have arisen in the late twentieth century and promise to accelerate in the future. In the era of big science, with systems of appointment, promotion, and rewards for scientists increasingly determined by external economic and political forces, it is doubtful that science is driven primarily by the simple quest for truth about nature for its own sake.

For all these reasons, there is today a pressing need for public as well as professional philosophical inquiry into the scientific enterprise, for discussion of the means, goals, costs, and risks as well as the benefits of scientific knowledge. Universal agreement will likely never be achieved on how to make value judgments regarding the conduct of science that would adjudicate all the conflicting interests typically involved. These questions can only serve as a guide to considerations that usually need to be taken into account in policy decisions about scientific and technological developments. Surely it would be desirable, taking a cue from a central virtue in the scientific tradition itself, if everyone affected by such decisions—the general public as well as scientists, technologists, industrialists, government officials, politicians—were to make their value assumptions explicit so that these could be challenged and debated in an open political process.

Suggestions for Further Reading

Antony, L., and C. Witt, eds. 1993. *A Mind of One's Own: Feminist Essays on Reason and Objectivity.*

Bloor, D. 1978. *Science and Social Imagery.*
Bronowski, J. 1965. *Science and Human Values.*
Feyerabend, P. K. 1978. *Science in a Free Society.*
Hardin, G. 1985. *Filters Against Folly.*
Lehrer, K., ed. 1987. *Science and Ethics.*
Lowrance, W. W. 1976. *Of Acceptable Risk: Science and the Determination of Safety.*
Maxwell, N. 1984. *From Knowledge to Wisdom.*
Passmore, J. 1978. *Science and Its Critics.*
Putnam, H. 1990. *Realism with a Human Face.*
Shrader-Frechette, K. S. 1985. *Risk Analysis and Scientific Method.*
Woolgar, S. 1988. *Knowledge and Reflexivity: New Frontiers in the Sociology of Knowledge.*

References

Antony, L., and C. Witt, eds. 1993. *A Mind of One's Own: Feminist Essays on Reason and Objectivity*. Westview Press.

Aposhian, H. V. 1970. "The Use of DNA for Gene Therapy—The Need, Experimental Approach, and Implications." *Perspectives in Biology and Medicine* 14:98–108.

Appleyard, B. 1992. *Understanding the Present: Science and the Soul of Modern Man*. Pan Books.

Aristotle. 1984. *The Complete Works of Aristotle*. Edited by J. Barnes. Princeton University Press.

Asimov, I. 1982. *Asimov's Biographical Encyclopedia of Science and Technology: The Lives and Achievements of 1510 Great Scientists from Ancient Times to the Present Chronologically Arranged*. 2d ed. Doubleday & Co.

Bacon, F. 1870. *The Works of Francis Bacon*. Edited by J. Spedding, R. L. Ellis, and D. D. Heath. Longman's Green.

Baier, K., and N. Rescher, eds. 1969. *Values and the Future*. The Free Press.

Barnes, B. 1985. *About Science*. Blackwell.

Belloc, H. 1931. *Essays of a Catholic Layman in England*. Sheed & Ward.

Ben-David, H. 1971. *The Scientist's Role in Society*. Prentice-Hall.

Bentham, J. 1962. *The Works of Jeremy Bentham*. Edited by J. Bowring. Russell & Russell.

Berg, P. et al. 1974. "Potential Biohazards of Recombinant DNA Molecules." *Science* 185:303.

Berg, P., and M. Singer. 1976. "Seeking Wisdom in Recombinant DNA Research." *Federation Proceedings* 35:2542–2543.

Berman, M. 1981. *The Reenchantment of the World*. Cornell University Press.

Bernal, J. D. 1965. *Science in History*. Hawthorne Books.

Bernstein, J. 1978. *Experiencing Science*. Basic Books.

———. 1992. "The Farm Hall Transcripts: The German Scientists and the Bomb." *The New York Review* August 13, 39:47–53.

———. 1993. *Cranks, Quacks, and the Cosmos: Writing on Science*. Basic Books.

Beveridge, W.I.B. 1957. *The Art of Scientific Investigation*. Vintage Books.

Beyerchen, A. D. 1977. *Scientists Under Hitler: Politics and the Third Reich*. Yale University Press.

Bhatia, S. 1992. "The Man Who Holds Iraq's Nuclear Secrets." *The Observer*, May 17.

Bishop, J. E. 1993. "Cold Fusion." *Popular Science* (August):47–51, 82.

Blackett, P. M. S. 1933. *The Craft of Experimental Physics*. Cambridge University Studies.

Blake, W. 1969. *Blake: Complete Writings*. Edited by G. Keynes. Oxford University Press.

Bloor, D. 1976. *Knowledge and Social Imagery*. Routledge & Kegan Paul.

Blumberg, A., and G. Owens. 1976. *Energy and Conflict: The Life and Times of Edward Teller*. G. P. Putnam's Sons.

Bowlby, J. 1990. *Charles Darwin: A New Life*. W. W. Norton.

Bradley, D. 1967. *Count Rumford*. Van Nostrand.

Brewster, D. 1965. *Memoirs of the Life, Writings, and Discoveries of Sir Isaac Newton*. Johnson Reprint Corporation.

Broad, W. J., and N. Wade. 1982. *Betrayers of the Truth*. Simon & Schuster.

Bronowski, J. 1965. *Science and Human Values*. Rev. ed. Harper & Row.

Burke, J. 1978. *Connections*. Little, Brown.

Burrt, E. A. 1932. *The Metaphysical Foundations of Modern Science*. Doubleday.

Burt, C. 1955. "The Evidence for the Concept of Intelligence." *British Journal of Educational Psychology* 25:158–177.

Bury, J. B. 1932. *The Idea of Progress: An Inquiry into Its Origin and Growth*. Macmillan.

Butterfield, H. 1957. *The Origins of Modern Science*. 2d ed. Macmillan.

Carson, R. 1962. *Silent Spring*. Houghton Mifflin.

Carter, R. 1965. *Breakthrough: The Saga of Jonas Salk*. Trident Press.

Cassidy, D. C. 1992. *Uncertainty: The Life and Science of Werner Heisenberg*. W. H. Freeman.

Cavalieri, L. F. 1981. *The Double-Edged Helix*. Columbia University Press.

Chain, E. B. 1970. "Social Responsibility and the Scientist." *New Scientist* 22:166–170.

Chalmers, A. F. 1982. *What Is This Thing Called Science?* 2d ed. University of Queensland Press.

———. 1990. *Science and Its Fabrication*. Open University Press.

Cherfas, J. 1982. *Man-Made Life: A Genetic Engineering Primer*. Blackwell.

Clark, R. W. 1971. *Einstein: the Life and Times*. Thomas Y. Crowell.

———. 1985. *The Life of Ernst Chain: Penicillin and Beyond*. Weidenfeld & Nicolson.

Clifford, W. K. 1898. *The Common Sense of the Exact Sciences*. Kegan Paul, Treich, Truber & Co.

Close, F. 1990. *Too Hot to Handle: The Race for Cold Fusion*. Princeton University Press.

Cohen, I. B. 1985a. *Revolution in Science*. Harvard University Press.

———. 1985b. *The Birth of a New Physics*. Rev. ed. W. W. Norton.

Cohen, R. 1974. "Ethics in Science." In *Science, Technology and Freedom*, edited by W. H. Truitt and T.W.G. Solomens. Houghton Mifflin.

Commoner, B. 1971a. *The Closing Circle: Nature, Man and Technology*. Alfred A. Knopf.

———. 1971b. "The Ecological Crisis." In *The Social Responsibility of the Scientist*, edited by M. Brown. The Free Press.

————. 1990. *Making Peace with the Planet.* Pantheon.

Conant, J. B. 1951. *On Understanding Science.* Yale University Press.

Copleston, F. 1960. *A History of Philosophy.* Vol. 6, part 1. Image Books.

Cornelius, D. K., and E. St. Vincent. 1964. *Cultures in Conflict: Perspectives on the Snow Leavis Controversy.* Scott, Foresman & Co.

Coulson, T. 1950. *Joseph Henry: His Life and Work.* Princeton University Press.

Cousins, N. 1979. "The Fallacy of Cost-Benefit Ratio." *Saturday Review* 6:8.

Crowther, J. G. 1941. *The Social Relations of Science.* Macmillan.

Cummings, E. E. 1991. *Complete Poems, 1904–1962.* Edited by G. T. Firmage. Liveright.

Curie, E. 1937. *Madame Curie.* Translated by V. Sheean. Doubleday & Co.

Darwin, F. 1899. *The Life and Letters of Charles Darwin.* D. Appleton.

Davies, P. 1993. "A Window into Science." *Natural History* 102 (July):68–71.

DeRopp, R. S. 1992. *The New Prometheans: Creative and Destructive Forces in Modern Science.* Dell.

Descartes, R. 1955. *The Philosophical Works of Descartes.* Translated by E. S. Haldane and G.R.T. Ross. Dover Publications.

Desmond, A., and J. Moore. 1991. *Darwin: The Life of a Tormented Evolutionist.* Warner Books.

d'Holbach, Baron. 1868. *The System of Nature; or the Laws of the Moral and Physical World.* Boston.

Dickenson, H. W. 1936. *James Watt.* Cambridge University Press.

Dixon, B. 1973. *What Is Science For?* Collins.

Dostoevsky, F. 1960. *Notes from Underground.* Edited and translated by R. E. Matlaw. E. P. Dutton & Co.

Drake, S. 1978. *Galileo at Work: His Scientific Biography.* University of Chicago Press.

Dubos, R. J. 1950. *Louis Pasteur.* Little, Brown.

————. 1961. *The Dreams of Reason.* Columbia University Press.

Easlea, B. 1983. *Fathering the Unthinkable: Masculinity, Scientists and the Nuclear Arms Race.* Pluto Press.

Edwards, P. ed. 1967. *The Encyclopedia of Philosophy.* 8 vols. Macmillan.

Ehrlich, P. 1968. *The Population Bomb.* Ballantine Books.

Ehrlich, P., and A. Ehrlich. 1990. *The Population Explosion.* Simon & Schuster.

Einstein, A. 1950. *Out of My Later Years.* Philosophical Library.

————. 1954. *Ideas and Opinions.* Crown.

Eisenstein, E. L. 1979. *The Printing Press as an Agent of Change.* Cambridge University Press.

Evlanoff, M., and M. Fluor. 1969. *Alfred Nobel: The Loneliest Millionaire.* The Ward Ritchie Press.

Farrington, B. 1949. *Greek Science: Its Meaning for Us.* Penguin Books.

————. 1966. *The Philosophy of Francis Bacon.* Phoenix Books.

————. 1969. *Francis Bacon: Pioneer of Planned Science.* Frederick A. Praeger.

Fauvel, J. et al., eds. 1988. *Let Newton Be!* Oxford University Press.

Feuer, L. S. 1992. *The Scientific Intellectual: The Psychological & Sociological Origins of Modern Science.* Transaction Publishers.

Feyerabend, P. K. 1975. "How to Defend Society Against Science." In *Scientific Revolutions*, edited by I. Hacking (1981). Oxford University Press.
————. 1978. *Science in a Free Society*. New Left Books.
————. 1988. *Against Method*. Rev. ed. Verso.
Feynman, R. P. 1958. "The Value of Science." In *Frontiers in Science*, edited by E. Hutchings. Basic Books.
Fisher, R. A. 1936. "Has Mendel's Work Been Rediscovered?" *Annals of Science* 1:115–137.
Florman, S. C. 1981. *Blaming Technology: The Irrational Search for Scapegoats*. St. Martin's Press.
Fuller, R. B. 1969. *Utopia or Oblivion: The Prospects for Humanity*. Bantam Books.
Galilei, Galileo. 1957. *Discoveries and Opinions of Galileo*. Translated by S. Drake. Doubleday Anchor Books.
Gardner, M. 1981. *Science: Good, Bad and Bogus*. Oxford University Press.
Gendron, B. 1977. *Technology and the Human Condition*. St. Martin's Press.
Gibbs, F. W. 1965. *Joseph Priestley: Adventures in Science and Champion of Truth*. T. Nelson.
Gillispie, C. C., ed. 1970–1980. *Dictionary of Scientific Biography*. 16 vols. Charles Scribner's Sons.
Gjertsen, D. 1989. *Science and Philosophy*. Penguin Books.
Goethe, J. W. 1840. *Theory of Colours*. Translated by C. L. Eastlake. John Murray.
Goodchild, P. 1980. *J. Robert Oppenheimer*. BBC.
Goran, M. 1967. *The Story of Fritz Haber*. University of Oklahoma Press.
Groves, L. R. 1963. *Now It Can Be Told*. Andre Deutsch.
Habermas, J. 1971. *Knowledge and Human Interests*. Beacon Press.
Hacking, I., ed. 1981. *Scientific Revolutions*. Oxford University Press.
————. 1983. *Representing and Intervening*. Cambridge University Press.
Hall, A. R. 1962. *The Scientific Revolution, 1500–1800*. Beacon Press.
Hall, S. S. 1987. *Invisible Frontiers: The Race to Synthesize a Human Gene*. Microsoft Press.
Hama, N. 1993. "The 21st Century from an Economic Perspective." In *Visions for the 21st Century*, edited by S. Moorcroft. Praeger.
Hanson, N. R. 1958. *Patterns of Discovery*. Cambridge University Press.
Hardin, G. 1968. "The Tragedy of the Commons." *Science* 162:1243–1248.
————. 1985. *Filters Against Folly*. Viking Penguin.
Harding, S. 1986. *The Science Question in Feminism*. Cornell University Press.
Harding, S., ed. 1993. *The "Racial" Economy of Science: Toward a Democratic Future*. Indiana University Press.
Hare, R. 1970. *The Birth of Penicillin*. George Allen & Unwin.
Harman, W. 1993. "The Second Scientific Revolution." In *Visions for the 21st Century*, edited by S. Moorcroft. Praeger.
Heilbron, J. 1986. *The Dilemmas of an Upright Man: Max Planck as a Spokesman for German Science*. University of California Press.
von Helmholtz, H. 1893. *Popular Lectures on Scientific Subjects*. 1st series. New York.

Hempel, C. G. 1960. "Science and Human Values." In *Social Control in a Free Society*, edited by R. E. Spiller. University of Pennsylvania Press.

————. 1966. *Philosophy of Natural Science*. Prentice-Hall.

Holton, G. 1986. *The Advancement of Science, and Its Burdens*. Cambridge University Press.

Huizinga, J. R. 1993. *Cold Fusion: The Scientific Fiasco of the Century*. New York.

Hull, D. 1988. *Science as a Process: An Evolutionary Account of the Social and Conceptual Development of Science*. University of Chicago Press.

————. 1989. *The Metaphysics of Evolution*. State University of New York Press.

Hume, D. 1854. *The History of England from the Invasion of Julius Caesar to the Abdication of James the Second, 1688*. Phillips Sampson.

Huxley, J. 1957. *Knowledge, Morality, and Destiny*. New American Library.

Huxley, L. ed. 1900. *Life and Letters of Henry Huxley*. Vols. 1 and 2. Macmillan.

Huxley, T. H. 1964. *Science and Education*. The Citadel Press.

Irvine, W. 1955. *Apes, Angels, and Victorians*. McGraw-Hill.

Jonas, H. 1984. *The Imperative of Responsibility: In Search of an Ethics for the Technological Age*. University of Chicago Press.

————. 1985. "Ethics and Biogenetic Art." In *Taking Sides: Clashing Views on Controversial Bioethical Issues*, 3d ed., edited by C. Levine. Duskin.

Joravsky, D. 1970. *The Lysenko Affair*. Harvard University Press.

Judson, H. F. 1980. *The Search for Solutions*. Hutchinson.

Juenger, F. G. 1956. *The Failure of Technology*. H. Regnery.

Jungk, R. 1958. *Brighter Than a Thousand Suns: A Personal History of the Atomic Scientists*. Harcourt Brace.

Kamen, M. D. 1984. *Radiant Science, Dark Politics: A Memoir of the Nuclear Age*. University of California Press.

Kamin, L. J. 1974. *The Science and Politics of I.Q.* Harmondsworth.

Kant, I. 1956. *Critique of Practical Reason*. Translated by L. W. Beck. The Bobbs-Merrill Company.

————. 1959. *Foundations of the Metaphysics of Morals*. Translated by L. W. Beck. Macmillan.

Kearney, H., ed. 1964. *Origins of the Scientific Revolution*. Longman's Green.

Keller, E. F. 1983. *A Feeling for the Organism*. W. H. Freeman.

Kenney, M. 1986. *Biotechnology: The University-Industrial Complex*. Yale University Press.

Kepler, J. 1981. *Mysterium Cosmographicum: The Secret of the Universe*. Translated by A. M. Duncan, commentary by E. S. Aiton. Abaris Books.

Kevles, D. J., and L. Hood, eds. 1992. *The Code of Codes: Science and Social Issues in the Human Genome Project*. Harvard University Press.

Keynes, G. 1966. *The Life of William Harvey*. Clarendon Press.

Kinsman, F. 1993. "Business to the Rescue." In *Visions for the 21st Century*, edited by S. Moorcroft. Praeger.

Klein, A. E. 1972. *Trial by Fury: The Polio Vaccine Controversy*. Charles Scribner's Sons.

Koestler, A. 1960. *The Watershed: A Biography of Johannes Kepler*. Doubleday & Co.

Koyré, A. 1973. *The Astronomical Revolution.* Methuen & Co.

Krimsky, S. 1982. *Genetic Alchemy: The Social History of the Recombinant DNA Controversy.* MIT Press.

Krutch, J. W. 1929. *The Modern Temper.* Harcourt, Brace & World.

Kuhn, T. S. 1962. *The Structure of Scientific Revolutions.* University of Chicago Press.

————. 1977. *The Essential Tension: Selected Studies in Scientific Tradition and Change.* University of Chicago Press.

La Farge, O. 1965. "Scientists are Lonely Men." In *The New Treasury of Science,* edited by Harlow Shapely et al. Harper & Row.

LaFollette, M. C. 1990. *Making Science Our Own: Public Images of Science 1910–1955.* University of Chicago Press.

Langone, J. 1978. *Human Engineering, Marvel or Menace?* Little, Brown and Company.

Latour, B., and S. Woolgar. 1979. *Laboratory Life: The Social Construction of Scientific Facts.* Sage Publications.

Lear, J. 1978. *Recombinant DNA, The Untold Story.* Crown.

Leavis, F. R. 1962. "Two Cultures? The Significance of C. P. Snow." In *An Essay on Sir Charles Snow's Rede Lecture,* edited by F. R. Leavis and M. Yudkin. Pantheon Books.

Lehrer, K., ed. 1987. *Science and Ethics.* Rodopi.

Leopold, A. 1949. *A Sand County Almanac.* Oxford University Press.

Letto, J. 1992. "One Hundred Years of Compromise." *Buzzworm* 4 (March/April):26–32.

Levins, R., and R. C. Lewontin. 1976. "The Problem of Lysenkoism." In *The Radicalisation of Science,* edited by H. Rose and S. Rose. Macmillan.

Lewontin, R. C. 1991. *Biology as Ideology: The Doctrine of DNA.* HarperCollins.

Lloyd, G.E.R. 1987. *The Revolution of Wisdom: Studies in the Claims and Practice of Ancient Greek Science.* University of California Press.

Losee, J. 1980. *A Historical Introduction to the Philosophy of Science.* 2d ed. Oxford University Press.

Love, A., and J. Childers, eds. 1965. *Listen to Leaders in Science.* Van Rees Press.

Lowrance, W. W. 1976. *Of Acceptable Risk: Science and the Determination of Safety.* William Kaufmann.

————. 1986. *Modern Science and Human Values.* Oxford University Press.

MacFarlane, G. 1984. *Alexander Fleming: The Man and the Myth.* Hogarth Press.

Mackay, A. L. 1991. *A Dictionary of Scientific Quotations.* Institute of Physics Publishing.

MacLeish, A. 1956. "Why Do We Teach Poetry?" *Atlantic Monthly* (March):48–53.

Marchant, J. 1916. *Alfred Russell Wallace: Letters and Reminiscences.* Harper & Brothers.

Marcuse, H. 1964. *One-Dimensional Man.* Routledge & Kegan Paul.

Marks, J. 1983. *Science and the Making of the Modern World.* Macmillan.

Marx, L. 1964. *The Machine in the Garden: Technology and the Pastoral Ideal in America.* Oxford University Press.

Mason, S. F. 1962. *A History of the Sciences.* Macmillan.

Maxwell, N. 1984. *From Knowledge to Wisdom.* Oxford University Press.

McCay, M. A. 1993. *Rachel Carson.* Twayne.

McKie, D. 1952. *Antoine Lavoisier: Scientist, Economist, Social Reformer.* Henry Schuman.

McLuhan, M. 1969. *The Gutenberg Galaxy.* Mentor Books.

McMurry, L. O. 1981. *George Washington Carver: Scientist and Symbol.* Oxford University Press.

Medawar, P. B. 1972. *The Hope of Progress.* Methuen.

————. 1984. *The Limits of Science.* Oxford University Press.

Medvedev, Z. 1969. *The Rise and Fall of T. D. Lysenko.* Columbia University Press.

Merchant, C. 1980. *The Death of Nature: Women, Ecology, and the Scientific Revolution.* Harper & Row.

————. 1992. *Radical Ecology: The Search for a Livable World.* Routledge.

Merton, R. K. 1957. *Social Theory and Social Structure.* Rev. ed. The Free Press.

Michelmore, P. 1969. *The Swift Years: The Robert Oppenheimer Story.* Dodd, Mead & Co.

Monod, J. 1971. *Chance and Necessity: An Essay on the Natural Philosophy of Modern Biology.* Knopf.

Moore, R. 1967. *Niels Bohr: The Man and the Scientist.* Hodder & Stroughton.

More, L. T. 1934. *Isaac Newton.* Charles Scribner's Sons.

Mott, N., and R. Peierls. 1977. "Werner Heisenberg." *Biographical Memoirs of Fellows of the Royal Society* 23:213–251.

Mueller-Hill, B. 1993. "Science, Truth and Other Values." *The Quarterly Review of Biology* 68:399–407.

Newman, J. R. 1961. *Science and Sensibility.* Simon & Schuster.

Newton, I. 1962. *Principia.* Vols. 1 and 2. Rev. ed. Edited by F. Cajori. University of California Press.

Newton-Smith, W. 1981. *The Rationality of Science.* Routledge and Kegan Paul.

Nicolson, M. 1956. *Science and Imagination.* Cornell University Press.

Olson, R. 1990. *Science Deified & Science Defied: The Historical Significance of Science in Western Culture.* Vols. 1 & 2. University of California Press.

Oppenheimer, J. R. 1986. "Physics in the Contemporary World." In *The Sacred Beetle and Other Great Essays in Science,* edited by M. Gardner, New American Library.

Otto, M. C. 1940. *The Human Enterprise.* F. S. Crofts.

Pais, A. 1991. *Niel Bohr's Times in Physics, Philosophy and Polity.* Oxford University Press.

Passmore, J. 1974. *Man's Responsibility for Nature: Ecological Problems and Western Traditions.* Charles Scribner's Sons.

————. 1978. *Science and Its Critics.*

Peirce, C. S. 1877. "The Fixation of Belief." *The Popular Science Monthly* 12:1–15.

Perutz, M. F. 1989. *Is Science Necessary? Essays on Science and Scientists.* Barrie and Jenkins.

Pirsig, R. 1974. *Zen and the Art of Motorcycle Maintenance: An Inquiry into Values.* Bantam Books.

Planck, M. 1936. *The Philosophy of Physics*. W. W. Norton.

Plato. 1961. *The Collected Dialogues of Plato*. Edited by E. Hamilton and H. Cairns. Bollingen Foundation.

Poincaré, H. 1952. *Science and Method*. Dover Publications.

Popper, K. R. 1963. *Conjectures and Refutations*. Harper & Row.

————. 1975. "The Rationality of Scientific Revolutions." In *Scientific Revolutions*, edited by H. Hacking. Oxford University Press.

Powers, T. 1993. *Heisenberg's Wars: The Secret History of the German Bomb*. Knopf.

Price, D. de Solla. 1962. *Science Since Babylon*. Yale University Press.

————. 1963. *Little Science, Big Science*. Columbia University Press.

Putnam, H. 1981. *Reason, Truth and History*. Cambridge University Press.

————. 1990. *Realism with a Human Face*. Harvard University Press.

Ravetz, J. 1971. *Scientific Knowledge and Its Social Problems*. Oxford University Press.

Redner, H. 1987. *The Ends of Science: An Essay in Scientific Authority*. Westview Press.

Reid, R. W. 1969. *Tongues of Conscience: War and the Scientist's Dilemma*. Constable.

Rhoads, S. E. ed. 1980. *Valuing Life: Public Policy Dilemmas*. Westview Press.

Rifkin, J. 1981. *Declaration of a Heretic*. Routledge & Kegan Paul.

————. 1991. *Biosphere Politics: A New Consciousness for a New Century*. Crown.

Root-Bernstein, R. 1988. *Discovering*. Harvard University Press.

Rossi, P. 1970. *Philosophy, Technology, and the Arts in the Early Modern Era*. Harper & Row.

Roszak, T. 1972. *Where the Wasteland Ends: Politics and Transcendence in Postindustrial Society*. Doubleday.

Russell, B. 1903. "A Free Man's Worship." *The Independent Review* (December). Reprinted in *Mysticism and Logic* (1917). Allen & Unwin.

————. 1945. *A History of Western Philosophy*. Simon and Schuster.

Sakharov, A. 1990. *Memoirs*. Knopf.

Salmon, W. C. 1975. *The Foundations of Scientific Inference*. Pittsburgh University Press.

Sarton, G. 1927–1948. *An Introduction to the History of Science*. 3 vols. Williams and Wilkins.

————. 1936. *The Study of the History of Science*. Dover Publications.

————. 1962. *The History of Science and the New Humanism*. Indiana University Press.

Sayre, A. 1975. *Rosalind Franklin and DNA*. W. W. Norton.

Schofield, R. E. 1966. *A Scientific Autobiography of Joseph Priestley, 1733–1804*. MIT Press.

Shapely, H. et al., eds. 1965. *The New Treasury of Science*. Harper & Row.

Shelley, M. W. 1994. *Frankenstein, or the Modern Prometheus*. Quality Paperback Book Club.

Shrader-Frechette, K.S. 1985. *Risk Analysis and Scientific Method*. Reidel.

Singer, C. 1958. *From Magic to Science*. Dover.

Sinsheimer, R. L. 1975. "Troubled Dawn for Genetic Engineering." *New Scientist* 68:148–151.

———. 1976. "On Coupling Inquiry and Wisdom." *Federation Proceedings* 35:2540–2542.

Snow, C. P. 1959. *The Two Cultures and the Scientific Revolution*. Cambridge University Press.

Sootin, H. 1959. *Gregor Mendel: Father of the Science of Genetics*. The Vanguard Press.

Sylvester, E. J., and L. C. Klotz. 1983. *The Gene Age*. Charles Scribner's Sons.

Teller, E. 1955. "The Work of Many People." *Science* (February)121:267–275.

———. 1945. *A History of Western Philosophy*. Simon and Schuster.

———. 1960. *The Reluctant Revolutionary*. University of Missouri Press.

Thomas, L. 1973. "Natural Man." In *The Lives of a Cell: Notes of a Biology Watcher*. Viking Penguin.

Thompson, D. 1989. "The Most Hated Man in Science." *Time*, December 4, 102–104.

Tobias, M. ed. 1985. *Deep Ecology*. Avant Books.

Trilling, L. 1962. "Science, Literature & Culture: A Comment on the Leavis-Snow Controversy." *Commentary* 33:461–477.

Turnbull, H. W. 1951. *The Great Mathematicians*. Methuen & Co.

Van der Waerden, B. L. 1961. *Science Awakening*. Oxford University Press.

Vico, G. 1982. *Vico: Selected Writings*. Edited and translated by L. Pompa. Cambridge University Press.

Wald, G. 1976. "The Case Against Genetic Engineering." *The Sciences* 16:7–10.

Watson, J. D. 1968. *The Double Helix: A Personal Account of the Discovery of the Structure of DNA*. Atheneum.

———. 1980. *The Double Helix*. Norton Critical Edition. Edited by G. S. Stent. W. W. Norton.

Watson, J. D., and R. M. Cook-Deegan. 1990. "The Human Genome Project and International Health." *Journal of the American Medical Association* 263:3322–3324.

Watson, J. D., and J. Tooze. 1981. *The DNA Story: A Documentary History of Gene Cloning*. W. H. Freeman.

Weaver, W. 1955. "Science and People." *Science* 122: 1255–1259.

Weinberg, S. 1977. *The First Three Minutes: A Modern View of the Origin of the Universe*. Basic Books.

Weintraub, S. 1963. *C. P. Snow: A Spectrum*. Charles Scribner's Sons.

Wernich, R. 1993. "The Worlds of Jefferson at Monticello." *Smithsonian* 24:80–93.

Westfall, R. S. 1980. *Never at Rest: A Biography of Isaac Newton*. Cambridge University Press.

Wheale, P., and R. McNally. 1988. *Genetic Engineering: Catastrophe or Utopia*. Harvester.

Wheeler, L. P. 1962. *Josiah Willard Gibbs*. Yale University Press.

White, M., and J. Gribbin. 1994. *Einstein, A Life in Science*. Dutton.

Whitehead, A. N. 1925. *Science and the Modern World.* Macmillan.

Wightman, W.P.D. 1950. *The Growth of Scientific Ideas.* Oliver & Boyd.

Wilson, C. 1972. "How Did Kepler Discover His First Two Laws?" *Scientific American* (March):92–106.

Woolgar, S. 1988. *Knowledge and Reflexivity: New Frontiers in the Sociology of Knowledge* . Sage.

Woolpart, L., and A. Richards. 1988. *A Passion for Science.* Oxford University Press.

Yoxen, E. 1983. *The Gene Business.* Pan Books.

Ziman, J. 1968. *Public Knowledge: The Social Dimension of Science.* Cambridge University Press.

About the Book and Authors

The development of modern science and its increasing impact on our life and culture is one of the great stories of our time. Coming to understand that story and coming to terms with the institution of modern science should be an important part of the education of every student.

In *The Many Faces of Science*, Leslie Stevenson and Henry Byerly masterfully and painlessly provide the basic information and the philosophical reflection students need to gain such understanding. Making good use of case study methods, the authors introduce us to dozens of figures from the history of science, highlighting both heroes and villains. Providing an elementary sketch of the development of science through the lives of its practitioners, Stevenson and Byerly bring the story alive through the examination of the often mixed motives of scientists, as well as the conflicting values people bring to science and to their perceptions of its impact on society. They also explore the relationship between scientific practice and political and economic power.

Brief, accessible, and rich in anecdotes, personal asides, and keen insight, *The Many Faces of Science* is the ideal interdisciplinary introduction for nonscientists in courses on science studies, science and society, and science and human values. It will also prove useful as supplementary reading in courses on science in philosophy, sociology, and political science.

Leslie Stevenson is reader in logic and metaphysics at the University of St. Andrews in Fife, Scotland. He is the author of *Seven Theories of Human Nature, The Metaphysics of Experience,* and many articles on language, mind, and science. **Henry Byerly** is professor of philosophy at the University of Arizona. He is the author of *A Primer of Logic* and has published many articles on biology as well as on the philosophy of science.

Index